Irregular Soldiers and Rebellious States

Irregular Soldiers and Rebellious States

Small-Scale U.S. Interventions Abroad

Michael P. Noonan

ROWMAN & LITTLEFIELD
Lanham • Boulder • New York • London

Published by Rowman & Littlefield
An imprint of The Rowman & Littlefield Publishing Group, Inc.
4501 Forbes Boulevard, Suite 200, Lanham, Maryland 20706
www.rowman.com

6 Tinworth Street, London SE11 5AL, United Kingdom

British Library Cataloguing in Publication Information Available

Library of Congress Cataloging-in-Publication Data
Names: Noonan, Michael P., author.
Title: Irregular soldiers and rebellious states : small-scale U.S. interventions abroad / Michael P. Noonan.
Other titles: Small-scale U.S. interventions abroad
Description: Lanham : Rowman & Littlefield Publishing Group, 2021. | Includes bibliographical references and index. | Summary: "Michael P. Noonan examines U.S. capabilities to conduct foreign internal defense and unconventional warfare. Using a newly developed typology and nine case studies, he places campaigns within broader contexts of military culture and history, offering 3 findings and 6 policy recommendations for considering when or not to use such capabilities"—Provided by publisher.
Identifiers: LCCN 2020052926 (print) | LCCN 2020052927 (ebook) | ISBN 9781442271296 (cloth) | ISBN 9781442271302 (paperback) | ISBN 9781442271319 (epub)
Subjects: LCSH: Irregular warfare—United States—Case studies. | Special forces (Military science)—United States—Case studies. | Military doctrine—United States.
Classification: LCC U240 .N66 2021 (print) | LCC U240 (ebook) | DDC 355.02/180973—dc23
LC record available at https://lccn.loc.gov/2020052926
LC ebook record available at https://lccn.loc.gov/2020052927

Contents

List of Tables and Figures

TABLES

FIGURES

Abbreviations

AFP	Armed Forces of the Philippines
AFRICOM	U.S. Africa Command
AID	Agency for International Development
AOR	Area of Responsibility
AQAM	al-Qaeda and Affiliated Movements
AQI	al-Qaeda in Iraq
AQIM	al-Qaeda in the Islamic Maghreb
ASG	Abu Sayyaf Group
BATT	British Army Training Team
CA	Civil Affairs
CCT	U.S. Air Force Combat Control Team
CIA	U.S. Central Intelligence Agency
CINC	Commander-in-Chief
COIN	Counterinsurgency
COM	Chief of Mission
CT	Counterterrorism
DLF	Dhofar Liberation Front
DOD	Department of Defense
DOS	Department of State
ESAF	El Salvador Armed Forces
EUCOM	U.S. European Command
FATA	Federally Administered Tribal Areas
FID	Foreign Internal Defense
FIS	Islamic Salvation Front
FMLN	Farabundo Marti National Liberation Front
FNLA	National Liberation Front of Angola
FY	Fiscal Year

GCC	Geographic Combatant Command
GDP	Gross Domestic Product
GEN	General
GIA	Armed Islamic Group
GPC	Great Power Competition
GPF	General Purpose Forces
GRP	Government of the Republic of the Philippines
GSPC	Salafist Group for Prayer and Combat
GWOT	Global War on Terrorism
HN	Host Nation
IC	Intelligence Community
IDAD	Internal Defense and Development
ISR	Intelligence, Surveillance, and Reconnaissance
JCET	Joint Combined Exchange Training
JSOC	Joint Special Operations Command
JSOTF	Joint Special Operations Task Force
JUSMAG	Joint U.S. Military Assistance Group
KDP	Kurdish Democratic Party
LIC	Low-Intensity Conflict
LRC	Light Reaction Company
LTG	Lieutenant General
MILF	Moro Islamic Liberation Front
MILGROUP	U.S. Military Group
MiTT	Military Transition Team
MOOTW	Military Operations Other Than War
MNLF	Moro National Liberation Front
MPLA	People's Movement for the Liberation of Angola
MTT	Mobile Training Team
NATO	North Atlantic Treaty Organization
NSDD	National Security Decision Directive
ODA	U.S. Army Special Forces Operational Detachment-A
OEF-A	Operation Enduring Freedom-Afghanistan
OEF-P	Operation Enduring Freedom-Philippines
OEF-TS	Operation Enduring Freedom-Trans Sahara
OPATT	Operational Planning and Assistance Training Team
OSS	Office of Strategic Services
PACOM	U.S. Pacific Command
PDRY	People's Democratic Republic of Yemen
PFLOAG	Popular Front for the Liberation of the Occupied Arabian Gulf
PN	Partner Nation
PSI	Pan Sahel Initiative
PUK	Patriotic Union of Kurdistan

SA	Security Assistance
SAD	CIA Special Activities Division
SAF	Sultan's Armed Forces
SAS	UK Special Air Service
SC	Security Cooperation
SEAL	U.S. Navy Sea, Air, Land commandos
SF	U.S. Army Special Forces
SFA	Security Force Assistance
SFAB	U.S. Army Security Force Assistance Brigade
SFG(A)	Special Forces Group (Airborne)
SMU	Special Mission Unit
SOCAFRICA	Special Operations Command, Africa
SOCEUR	Special Operations Command, Europe
SOCOM	U.S. Special Operations Command
SOCPAC	Special Operations Command, Pacific
SOF	Special Operations Forces
SOUTHCOM	U.S. Southern Command
TCAV	terrorist coordination assistance visit
TSC	Theater Security Cooperation
TSCTI	Trans-Saharan Counterterrorism Initiative
TSCTP	Trans-Sahara Counterterrorism Partnership
TTP	Tactics, Techniques, and Procedures
UNSC	United Nation Security Council
USG	U.S. Government
UNITA	National Union for the Total Independence of Angola
UW	Unconventional Warfare
WMD	Weapons of Mass Destruction

Acknowledgments

The impetus for this book comes from over thirty years of interest in the related topics. While I do not have a Special Operations Forces (SOF) background, I did serve on a Military Transition Team in Iraq in western Nineveh, Iraq, serving as an advisor with an Iraqi Army infantry battalion in 2006–2007. This experience led me to observe advisory activities up close and personally and see some of the strengths and weaknesses of working with a host nation (HN) force. Of course, the circumstances that led to large-scale operations in Afghanistan and Iraq are different in nature than those that will be explored here.

Another experience that sparked my interest in this subject was coordinating with a team that helped to develop a course for U.S. Army Special Forces (SF) personnel on network operations. Working with that team, many of whose members were SF veterans, allowed me to have numerous conversations on topics related to this book.

One more key shaping experience for my thinking on this book was my time getting to know the late Sam C. Sarkesian while a graduate student at Loyola University Chicago. Professor Sarkesian was a true soldier-scholar who wrote broadly on SOF, U.S. national security policy, and civil-military relations. He was also a retired Army lieutenant colonel, who was an early member of Colonel Aaron Banks 10th Special Forces Group (Airborne) from where he deployed to fight in the Korean War with the United Nations Partisan Infantry Korea.

Over the years I have come into contact with many individuals who have helped to shape my thinking on subjects both close to, and tangential to, this book. In particular, I would like to thank Frank Hoffman, David Maxwell, Ian Rice, Anna Simons, Aaron Stein, John Allen "Jay" Williams, and an anonymous senior SOF officer. Others whom I am particularly grateful to are

David Betz, Michael Boyle, Joseph Celeski, Stephen Cimbala, Colin Clarke, Anthony Cormack, Ryan Evans, Ronald Granieri, Stephen M. Grenier, Sam Helfont, John Hillen, Michael C. Horowitz, Robert D. Kaplan, Mark R. Lewis, Walter A. McDougall, Barak Mendelsohn, Robert Murray, John Nagl, Sean Naylor, Carl Prine, Peter Schraeder, Kalev "Gunner" Sepp, Ashley Stocker, Dominic Tierney, Clint Watts, and the students of the Defense Analysis Department at the Naval Postgraduate School that I was able to visit and talk to a few years ago. Any errors in fact or judgment in the book are mine alone.

At the Foreign Policy Research Institute, I would like to thank Rollie Flynn, Alan Luxenberg, and the late Harvey Sicherman for their guidance and support. One of the best parts of my work at the Foreign Policy Research Institute has been working with interns. For their efforts in helping with pieces of this book, I would like to specifically thank Samuel Byers, Charles Edwards-Kuhn, Ive Gonzalez, and Dylan Yachyshen. I would also like to thank Ahaan Bagwe, Otis Baker, Alexandre Caillot, Brian Gallagher, Gordon Liu, Julia Maenza, Grady MacPhee, Lindsay Meyerson, Heewon Park, Ian Pollack, Claire Redfield, George Romer, Britnee Smith, and Stephanie Vladimirova for their research assistance efforts over the years.

Special thanks go to the Smith Richardson Foundation for the generous support that made this book possible. I am especially grateful to Christopher Griffin, Kathy Lavery, Nadia Schadlow, and Marin Strmecki. Smith Richardson Foundation has been one of the few stalwart supporters of research on foreign and defense policy even during times when other foundations have pulled back in those areas.

I would also like to thank to Jonathan Sisk and the editors and staff at Rowman & Littlefield for all of their efforts in making this book a reality.

Last, but certainly not the least, I would like to thank my wife Jo for her patience and love. This book would not have been possible without her. I also would be remiss if I did not thank Oscar, Raven, Chase, Rocky, Marble, and Milo for their companionship over the years.

Prologue: Irregular Soldiers and Rebellious States

The attacks on September 11, 2001, ushered in a period of protracted counterterrorism (CT) and counterinsurgency (COIN) operations that the United States remains involved in to this day. This has caused some to question how, when, and why the United States should use the military elements of its power abroad. As former secretary of defense Robert M. Gates, himself no dove, told the cadets at the U.S. Military Academy at West Point in 2011, "But in my opinion, any future defense secretary who advises the president to again send a big American land army into Asia or into the Middle East or Africa should 'have his head examined,' as General MacArthur so delicately put it."[1] Regardless of one's views on the wars in Afghanistan and Iraq, it is quite clear that the foreign policy of the United States will continue to employ its armed forces for myriad purposes (from purely humanitarian purposes to large-scale combat operations). The nation still retains a global footprint of bases, operations, interests, and allies. This will not change in all but the degree of scale. The question, however, that will remain is how will the country use its vaunted military capabilities in a world that is much less triumphalist for the United States than where it stood in the post–Cold War short-lived era of the "Unipolar Moment."[2]

The election of President Donald J. Trump brought about at least a rhetorical recalibration of the global interests and actions of the United States. While the administration shifted on policy issues relating to the objectives of American foreign policy, with many changes in key personnel, it talked about the rehoning of the nation's military edge with robust defense budgets while also talking about pulling back the deployment of military forces overseas. Although this may have been seen as contradictory by some, it seems to have been quite natural for the zeitgeist of a consensus of some corners of American politics that feels that the United States should no longer bear the burden of being the global policeman.

At the same time, there were tensions in the Trump administration's direction. The president wanted to disengage from the Middle East, and yet we are still involved in CT operations from Syria down to Yemen. The president also wanted more fruitful engagement with Russia, and yet Russia remains engaged against the United States and its allies from Syria to the Balkans to the Baltic. China is the only area where the president seemed to be rhetorically tough, but that was largely in terms of the trade relations between the two global giants.

The Pentagon under Trump, for its part, called for a return of attention to great power competition (GPC) in U.S. defense strategy. This is taken from the small declassified snippet of the National Defense Strategy, which is the Defense Department's strategic guidance that is derived from the National Security Strategy. In some ways, this followed an earlier attempt by the Obama administration to pivot American national security policy to East Asia. In terms of defense strategy, this is a somewhat seductive argument that the United States can step away from the inherent messiness of COIN and nation-building "wars amongst the peoples"[3] and reengage with fighting capital-intensive conventional, large-scale wars if and when needed. Of course, as the strategic truism goes, the enemy—or enemies—get a vote, too.

The United States continues to face threats and challenges today that will likely try to avoid tangling with U.S. conventional warfighting capabilities. After nearly twenty years of war in Afghanistan and many years of war in Iraq, adversaries and competitors know that the United States still has difficulties in operating in "small wars" spaces where local political considerations are extremely important to the successful conduct of military operations. Adversaries such as the Islamic State and al-Qaeda and their offshoots know this. Their rage against the international order has not gone away. Similarly, strategic competitors, such as China and Russia, have been watching the geostrategic environment of the past almost two decades and have clearly drawn their own conclusions about American military power. Other challengers, such as Iran and North Korea, also have their own assumptions.

The United States today faces a disordered world where traditional military solutions will be only one tool among many to cope with complex problems.[4] The United States will still have myriad interests in regions of the world where regular and irregular threats posed by state and nonstate actors (e.g., international terrorists, narcotic traffickers, organized criminal networks, and terror-crime pipelines) drive political and security considerations affecting its national security interests. Furthermore, A. Wes Mitchell and Jakub Grygiel have noted:

> Amid the now globally accepted thesis of American decline, America's global rivals are doing what aspirant powers have done at moments of transition for millennia: hypothesis-testing. They are probing the top state on the outer limits of its power commitments, where its strategic appendages are most vulnerable and its strength is most thinly spread. If history is any guide at all, they are reading America's responses to gauge how much latitude they have to make low-cost revisions to the system in their favor. But both they and American allies are watching not just how America responds to probes in their own neck of the woods but also to the probes of powers—and to the needs of similarly situated allies—in other regions. Lacking the geopolitical equivalent of a stock market, they are gathering valuable cues about America's intentions in their own neighborhood by tracking how it handles revisionists at other points on the U.S. strategic perimeter.[5]

Many times, it will likely be in the U.S. interest to confront these threats indirectly. This is the case because, even for a superpower with global reach, our capabilities are finite and must be strategically decided upon how, where, and when to allocate these resources.

What does this mean for the United States? The argument advanced in this book is that when necessary the United States must be prepared to be able to support allies and friends overseas by helping such governments to *defend* themselves from "subversion, lawlessness, insurgency, violent extremism, terrorism, and other threats to its security."[6] In U.S. military doctrine, this is referred to as foreign internal defense (FID). On the flipside, it must also be able to conduct *offensive* "operations and activities that are conducted to enable a resistance movement or insurgency to coerce, disrupt, or overthrow a government or occupying power by operating through or with an underground, auxiliary, and guerrilla force in a denied area."[7] This is called unconventional warfare (UW). Both FID and UW are not solely undertaken by military forces—they might involve many actors, such as people from the intelligence community (IC) or development personnel. These types of missions are sometimes referred to using the shorthand "indirect approaches."

In the United States, military units assigned to special operations forces (SOF) are often those used in order to indirectly achieve the goals of American foreign and defense policy—usually by working through, with, or by a proxy. The 2012 Defense Policy Guidance, for instance, noted:

> Across the globe we will seek to be the security partner of choice, pursuing new partnerships with a growing number of nations including those in Africa and Latin America—whose interests and viewpoints are merging into a common vision of freedom, stability, and prosperity. *Whenever possible, we will develop innovative, low-cost, and small-footprint approaches to achieve our security objectives*, relying on exercises, rotational presence, and advisory capabilities.[8]

Following upon this, the 2018 National Defense Strategy discusses the Global Operating Model—"how the Joint Force will be postured and employed to achieve its competition and wartime missions"—and states that one of its four layers, contact, will "help us compete more effectively below the level of armed conflict."[9] This attention has focused on SOF and SOF-like capabilities (e.g., Security Force Assistance Brigades and Special Marine Air-Ground Task Forces), and in particular using smaller-scale advisory missions, to build partnership capacity among allies, friends, and other states occupying key strategic geography. But such indirect approaches are not limited to purely defensive operations supporting other states to defend themselves; there are also offensive uses of indirect approaches such as working with the Northern Alliance in Afghanistan in 2001 or with the Peshmerga in northern Iraq in 2003.

But what is meant by indirect approaches? Many associate the phrase "indirect approach" with the writings of Basil Liddell Hart. For Liddell Hart, the indirect approach is about unbalancing one's enemies rather than going at them directly.[10] More recently, however, the phrase tends to be used in discussing the small-scale employment of SOF and other actors in complex political-military operational environments. SF lieutenant colonel Fernando M. Lujan, for instance, calls this approach "light footprint" and has argued as follows:

> Instead of "nation building" with large, traditional military formations, civilian policymakers are increasingly opting for a combination of air power, special operators, intelligence agents, indigenous armed groups and contractors, often leveraging relationships with allies and enabling partner militaries to take more active roles.[11]

But others see SOF as being divided between "shooters" and "advisors" or between "surgical strike" and "special warfare" elements that focus on different roles and missions.

This study defines indirect approaches as the small-scale employment of military forces and other actors (e.g., diplomats, intelligence officers, and law enforcement types) in times of either peace or war—which in itself has become a false dichotomy—in order to help other nations to defend themselves or to weaken or take down other formal or informal government structures abroad. (Such approaches are preferably used alone, but may in some circumstances be in concert with more direct applications of force.) To facilitate analysis, I develop a typology of offensive and defensive applications of indirect approaches through an examination of recent historical cases of the United States. This can help shed light on significant policy questions such as the following: Are there certain characteristics or conditions that will allow an indirect approach more likely to work? What types of forces should be used to carry out such operations? And what are the civil-military implications from these types of operations?

***

Retired SF colonel and historian John Waghelstein has noted that the U.S. military has historically had a bias against the types of forces trained for the indirect employment of force. In his words, "Beginning with General George Washington, who had a notorious disdain for irregular forces and partisan operations, the institutional Army has in the main tended to regard small wars as distractions from the main task of preparing to fight great conventional engagements."[12] But starting with World War II, the United States began to expand its use of indirect approaches to serve as an adjunct to its more direct forms of military power.

At that time, the indirect approach was seen as an adjunct form of military power using specially trained forces either to lend support to the operations of conventional forces or else as force multipliers in the so-called economy of force theaters.[13] The Office of Strategic Services (OSS), for instance, helped to raise or assist partisan forces to help fight in the war effort against the Axis powers. But the end of that war did not obviate the need for using indirect approaches. The rise of the Soviet threat and the geopolitical chess match for control over territory and peoples caused a rise in proxy warfare competition.

The OSS morphed into the Central Intelligence Agency (CIA) and the new Pentagon began to build—with many growing pains—irregular warfare capabilities to help counter the Soviets. SF were founded in 1952, for instance, with the idea of creating a force capable of building partisan forces to fight the Soviets behind their lines. They were first used in the Korean War. Other military officers and personnel were used to serve as advisors in places such as Greece and the Philippines to help fight against Soviet-backed subversion throughout the 1950s.

By the early 1960s, President John F. Kennedy was convinced that the threat posed by the Soviet Union needed expanded capabilities of both offensive and defensive nature to counter them. The army's Special Forces would now become experts at COIN, and other units, such as the Navy's Sea, Air, Land commandos (SEALs), would be formed. The presence of American advisors would expand, particularly in Southeast Asia.

The Vietnam War and its aftermath would cause some to question the efficacy of such advisors and forces, but the Soviet Union and its aggressive actions in the Third World continued. The United States responded with both covert and noncovert means in such places as Angola, Afghanistan, El Salvador, and Nicaragua until the end of the Cold War. At the same time, the military also developed "hyperconventional" elite units, such as the Rangers and 1st Special Forces Operational Detachment-Delta (Delta Force), in order to handle new missions, such as CT in the wake of the explosion of international terrorist attacks in the 1970s.[14] (Of course, to be fair, there may be a subtle shift in this area, Delta units were said to be deeply involved with UW efforts in Syria.[15])

The end of the Cold War replaced the overarching struggle with the Soviet Union with an underarching struggle against more diffuse threats.[16] Countering the inflow of narcotics to the United States and fighting terrorism were two such areas of concerns. The attacks on 9/11 by al-Qaeda, however, changed that dynamic. While Afghanistan and Iraq displayed large-scale uses of force, there were also numerous smaller-scale operations. This was only natural because al-Qaeda and affiliated movements (AQAM) was a dispersed network, and the United States had to work with numerous states to confront the threat.

In order to get a handle on this threat in the decade post-9/11, various "train and equip" authorities were authorized by Congress to allow the military the flexibility to build such partnership capacity abroad to take the fight to AQAM. Chief among these authorities were the so-called 1206 and 1208 provisions. Named for their paragraph numbers in National Defense Authorization Acts, the 1206, later named 1204, authority enabled U.S. forces to train and equip friendly militaries and the 1208 authority, later named section 1203 and now codified as Section 127e,[17] allows U.S. forces to train and equip foreign security forces or individuals for CT activities. (But some feel that such blanket authorities constrain our thinking rather than requesting specific authorities based upon needs driven by campaign plans designed to handle operational situations on the ground.[18] Others, however, feel that if you do not get authorities before the crisis, you will never get them during it and then wonder why certain capabilities are not being used.[19])

Special Operations Command (SOCOM), for its part, since the start of the Global War on Terrorism (GWOT), has tried, largely unsuccessfully, to increase its authorities and flexibility to conduct global operations with a minimum of hindrance from bureaucratic seams and impediments. A former SOCOM commander, retired admiral William McRaven, for instance, repeatedly talked about having more authority over the Combatant Command (COCOM) Theater Special Operations Commands (TSOCs) and for building a Global Special Operations Network. But no one questions the centrality of SOCOM's role in future defense strategy.

But such approaches are not limited solely to the SOF community. The U.S. Army, for instance, has begun to embrace part of the building-partnership capacity mission through elements of its Security Force Assistance Brigades (SFABs). The Air Force, too, has built some advisory structures to work in helping to build rotary and fixed wing capabilities with friends abroad. It remains to be seen what the full impact of these programs will be or whether, perhaps, they are initiatives driven purely to attract or retain resources when defense budgets were shrinking in the Obama administration.

From the previous examples, one can broadly sense that there are both offensive and defensive applications of such indirect approaches. Offensively, one immediately thinks of working with the Northern Alliance post-9/11 to take down the Taliban regime or perhaps the use of SOF to target key al-Qaeda leaders in such places as Somalia and Yemen. Defensive applications of the indirect approach could be seen over the past decade-plus in working with the governments of Colombia, Georgia, the Philippines, and of the nations of the Trans-Sahel in fighting al-Qaeda and Islamic State–affiliated groups.

What is different today, at least under current budgetary and strategic reality, is the centrality that some see in elevating such indirect approaches to the same level of importance as more direct approaches of military power. For instance, Navy SEAL Captain Rob Newson has argued that there is now a "MINFORCE" approach and states that such efforts

> should not be confused with an economy of force approach; designed as a holding action while the majority of available forces remain engaged in another, higher priority fight. This economy of force construct may have been the case as the wars raged in Iraq and Afghanistan. Today and in to the future, these small, distributed engagements with our partner nations should now be seen as the main effort. This is a new mental model, where size doesn't equate to priority, where distributed and remote nodes of our friendly network require innovative and agile staffing, logistics, fiscal, and intelligence support disproportionate to the numbers of personnel on the ground; where getting power to the edge of our network is fundamental to success against a distributed, adaptive, and resilient enemy network.[20]

If this is the case, then what we are seeing is a big shift in America's strategic and military culture(s). While Linda Robinson[21] and many others have written of building-partnership capacity and on disparate aspects of indirect approaches, no book puts all of these things together in one place to explain the possibilities and pitfalls of such approaches in an accessible way for both policymakers and the interested public. Understanding this with greater detail should ensure better applications of such approaches in the future as well as be instructive for when they should be avoided.

Much of the research conducted on indirect approaches is episodic and anecdotal; what is missing is a sustained analytical approach. We propose to fit the issue of the use of such indirect approaches within a typology of indirect interventions abroad. Such interventions will be examined across the axes of whether they are offensive (i.e., against a state or within the boundaries of a state) or defensive (i.e., in support of a state) and whether they

Table 0.1. Forms of Indirect Interventions

	Offensive	*Defensive*
Limited	Angola (1975–1980s), Afghanistan (1979–1989), Nicaragua (1981–1991)	El Salvador (1981–1992), Plan Colombia (1999–2015), Philippines (2001–2015), Georgia Train and Equip (2002–2004), Trans Sahara (2003–present), Central Africa (2011–2017)
Assertive	Afghanistan (2001–2002), northern Iraq (2002–2003)	Dhofar (1965–1975) [non-U.S. case]

are limited (i.e., the use of U.S. forces in combat operations is restricted) or assertive (i.e., U.S. combat involvement is relatively unrestricted). Table 0.1 graphically represents this typology and offers examples of these four types of indirect intervention.

Each of the possibilities of this four block schema will be fleshed out using the aforementioned cases. (Dhofar, while a non-U.S. case study, has been selected because it illustrates a type of indirect intervention that the U.S. has not been engaged in.) Only post-Vietnam cases have been selected in order to be able to make useful cross-case comparisons regarding issues such as governmental structure and the contextual matters. The four forms of indirect interventions will be examined using theory, history, and U.S. doctrine—with a particular emphasis on UW and FID. Within each of these categories of interventions, common characteristics will be examined to see what worked and did not work, how effectiveness was helped or hindered by the strategic context, the role played by personalities, and how unit type and the training of their personnel affected the situation on the ground. Furthermore, the issue of indirect interventions will be examined within the larger context of American strategic and military cultures.

It is important to know when to use—and when not to use—indirect approaches abroad in order to achieve the objectives of the United States at the appropriate cost in terms of blood, treasure, and prestige. Getting this right is essential, because as the military analyst Richard Gabriel pointed out in his book *Military Incompetence*:

> The ability of a nation to work its will through military force is one of the vital elements of national power. The credibility of a state's military forces (based on its record) is an important factor in its ability to gain its foreign-policy objectives *without* the use of force. The threat of force is often sufficient to gain policy objectives. . . . A pattern of repeated military failure can prompt or encourage an adversary to embark upon policies designed to take advantage of this pattern.[22]

Moving forward, offensive and defensive indirect approaches—whether they be limited or assertive—will be necessary to keep pressure on al-Qaeda and Islamic State–affiliated groups particularly in light of the foreign fighter problem emanating from the Afghanistan and Iraq wars and the current battlefields of Syria, among other locales. Terror-crime networks and other nonstate, substate, and state-based actors will also cause problems for countries sitting on geopolitically sensitive terrain where the United States has myriad interests.

In addition, this is an important matter because even if one believes that the United States has reentered an era of great power politics in East Asia or Eurasia the use of such indirect approaches will be important. Their use will be important because (1) there very well could be a rise in the conduct of proxy wars to indirectly achieve national interests where such forces would be vital in the execution of U.S. policy[23] and (2) many nations in East Asia, or to a lesser extent Eurasia, may not want a large presence of U.S. forces in order not to offend China, or Russia, but will want smaller-scale training assistance and engagement. Furthermore, even if there was an outbreak of large-scale conflict, indirect approaches would be used as a complement to more direct approaches.

The political will in the United States for large-scale uses of force abroad will be—absent a major attack—lacking. Options between committing 140,000 personnel on the ground, for instance, and doing nothing will be needed. This book offers examples for the tools the United States has at its disposal to conduct small-scale indirect interventions for offensive or defensive purposes that are either limited or assertive in nature. Of course, such tools are generally outside the mainstream of the U.S. national strategic and military cultures. Chapter 1 discusses these cultures, the history of U.S. military interventions, and the contemporary geostrategic environment confronting the United States. Chapter 2 discusses the nature of military advising in small-scale interventions. Defensive interventions are explored in chapter 3. Chapter 4 discusses offensive varieties of these interventions. Last, chapter 5 discusses findings and recommendations to improve U.S. performance when such small-scale interventions are required. While this book will be of interest to those with a deeper understanding of these topics, it is primarily meant as a primer for audiences interested in learning more about such indirect interventions.

NOTES

1. As Delivered by Secretary of Defense Robert M. Gates, "Speech," United States Military Academy, West Point, NY, Friday, February 25, 2011. Available at: http://www.defense.gov/speeches/speech.aspx?speechid=1539.

2. See, for example, Charles Krauthammer, "The Unipolar Moment," *Foreign Affairs*, America and the World 1990 Issue, pp. 23–33.

3. Rupert Smith, *The Utility of Force: The Art of War in the Modern World* (New York: Penguin Books, 2005).

4. See, for example, Nadia Schadlow, "Competitive Engagement: Upgrading America's Influence," *Orbis* 57, no. 4 (Autumn 2013): 501–515.

5. A. Wes Mitchell and Jakub Grygiel, "The Vulnerability of Peripheries," *The American Interest* 6, no. 4 (March/April 2011): 6.

6. Available at: https://www.jcs.mil/Portals/36/Documents/Doctrine/pubs/jp3_22.pdf?ver=2018-10-10-112450-103.

7. Available at: https://www.jcs.mil/Portals/36/Documents/Doctrine/pubs/jp3_05.pdf.

8. *Sustaining U.S. Global Leadership: Priorities for 21st Century Defense* (Washington, DC: Department of Defense, January 2012), italics in original. Available at: http://www.defense.gov/news/defense_strategic_guidance.pdf.

9. "*Global Operating Model.* The Global Operating Model describes how the Joint Force will be postured and employed to achieve its competition and wartime missions. Foundational capabilities include: nuclear; cyber; space; C4ISR; strategic mobility, and counter WMD proliferation. It comprises four layers: contact, blunt, surge, and homeland. These are, respectively, designed to help us compete more effectively below the level of armed conflict; delay, degrade, or deny adversary aggression; surge war-winning forces and manage conflict escalation; and defend the U.S. homeland." *Summary of the National Defense Strategy: Sharpening the American Military's Competitive Edge* (Washington, DC: Department of Defense, 2018), 7. Available at: https://dod.defense.gov/Portals/1/Documents/pubs/2018-National-Defense-Strategy-Summary.pdf.

10. See, for instance, B. H. Liddell Hart, *Strategy* (London: Faber and Faber Ltd., 1954).

11. Major Fernando M. Lujan, *Light Footprints: The Future of American Military Intervention* (Washington, DC: Center for a New American Security, March 2013), 5. He goes on to say: "The specific tactics involved in these operations may vary, but the guiding philosophy is clear: Send tens or hundreds instead of surging thousands. Be patient and work quietly within the constraints of the existing political and social ecosystem. Help others to help themselves instead of doing the work alone. But when necessary, act unilaterally with lethal, surgical precision" (p. 7).

12. Colonel (ret.) John D. Waghelstein, "What's Wrong in Iraq? Or Ruminations of a Pachyderm," *Military Review* 86, no. 1 (January–February 2006): 112.

13. DoD defines economy of force as follows: "The judicious employment and distribution of forces so as to expend the minimum essential combat power on secondary efforts in order to allocate the maximum possible combat power on primary efforts." Available at: http://www.dtic.mil/doctrine/dod_dictionary/data/e/22320.html.

14. See, for example: Sam C. Sarkesian, *The New Battlefield: The United States and Unconventional Conflicts* (Westport, CT: Praeger, 1986); Steven P. Basilici and Jeremy Simmons, "Transformation: A Bold Case for Unconventional Warfare" (MA Thesis, Naval Postgraduate School, June 2004); and Hy Rothstein, *Afghanistan and*

the Troubled Future of Unconventional Warfare (Annapolis, MD: Naval Institute Press, 2006).

15. Author conversation with a senior SF officer, April 2020.

16. Harvey Sicherman, "The Revenge of Geopolitics," *Orbis* 41, no. 1 (Winter 1997): 7–13.

17. 10 U.S. Code § 127e. Support of special operations to combat terrorism. Available at: https://www.law.cornell.edu/uscode/text/10/127e.

18. This point was raised by Colonel (ret.) David Maxwell in email correspondence on May 30, 2019.

19. Author conversation with a senior Army SOF officer, April 2020.

20. Rob Newson, "Adapting for the 'Other' War," *Small Wars Journal* (October 18, 2013). Available at: http://smallwarsjournal.com/jrnl/art/adapting-for-the-"other"-war.

21. See, for example, Linda Robinson, *The Future of U.S. Special Operations Forces* (New York: Council Special Report No. 66, Council on Foreign Relations, April 2013). Available at: http://www.cfr.org/special-operations/future-us-special-operations-forces/p30323?co=C009601.

22. Richard A. Gabriel, *Military Incompetence: Why the American Military Doesn't Win* (New York: Hill and Wang, 1985), 3.

23. See, for example, Andrew Mumford, *Proxy Warfare* (New York: Polity, 2013); and Tyrone Groh, *Proxy War: The Least Bad Option* (Stanford, CA: Stanford University Press, 2019).

Chapter 1

Uses of Force, the American Ways of War, and a Disordered World

Before jumping into a full discussion of U.S. small-scale interventions, it is important to establish the ways that the U.S. military has used its military power historically and the ways the national political and military cultures have thought about such applications of force. The dominant cultures, however, create friction for the modes of political-military thought needed to confront threats and challengers. Next, it is important to discuss the contemporary geostrategic environment. The contention here is that the present "disordered world" that the U.S. confronts will require more, rather than fewer, defensive and offensive indirect interventions.

Power and culture are two key factors affecting the behavior of nations in the international system. Power, which is fungible and can be shaped by environmental considerations such as natural resources, climate, and geography, helps determine the *means* at a state's disposal be they diplomatic, informational, military, or economic—the so-called DIME elements of power. Culture provides context[1] and shapes the *ways* that help to determine how a state exercises its power to achieve its *ends*. But culture is not always a monolithic factor within a state. Nation states sometimes are said to have a national strategic culture—or characteristics—that shape their strategic interactions with other actors in the international system.[2] However, below the national level, various institutions and other actors within a state can also have their own operating styles or cultures that shape the ways in which they would prefer to act or do react to events.[3]

While this book is focused upon looking at smaller-scale applications of American military and paramilitary power, it is important to place these sorts of uses of force within the broader aperture of American history. Here in this chapter we are interested in the interplay of the factors discussed earlier and how they have shaped the exercise of American military power abroad from the Founding to the present.

AMERICAN POWER AND THE AMERICAN WAYS OF WAR

The rise of American power was heavily influenced by the nation's unique form of government and the geopolitical realities it faced. Emerging from a protracted struggle for independence from the British Empire, the young nation was surrounded by rivals in the New World and had a weak central government. As the country matured and transitioned from the Articles of Confederation to the Constitution and as its economy and population grew the nation expanded—sometimes carefully and sometimes clumsily—westward and to the south. This expansion was not always peaceful.[4] And as the nation grew, so did its stature in international politics. Author Max Boot, for instance, argues that the United States has progressed through three eras in its growth of national power: (1) Commercial Power (1783–1898), (2) Great Power (1898–1941), and (3) Superpower (1941–present).[5] Similarly, the diplomatic historian Walter McDougall has, in terms of the nation's foreign policy, referred to Boot's commercial power era as the U.S. "promised land" period and post-1898 as its "crusader state" epoch.[6]

Culturally, representative democracy in the United States freed, at least theoretically, from the effects of the Old World nobility systems and the right to worship as one chose also added to the new nation's mystique. Westward and southern expansion took into account not only demographic growth and the quest for economic gain but also from a sense that democracy itself could not be contained. As McDougall puts it, such "expansion derived from the primordial, exceptional American commitment to liberty. Without freedom to grow, the nation would not be free at all."[7] Furthermore, overseas commerce brought the young nation into contact with nations and peoples farther and farther away from the North American continent.

While it has been argued that colonial America developed and possessed a way of war that was more attuned to the definition of small wars as used in this book (of a particular savage form that included the targeting of nonmilitary forces),[8] the experience of the Revolutionary War according to some solidified notions of U.S. military forces operating with more traditional forms of organizations and tactics. Trumpeting the late military historian Russell Weigley's claims that the American military has never been comfortable with guerrilla warfare, the retired SF colonel John Waghelstein has noted that "beginning with General George Washington, who had a notorious disdain for irregular forces and partisan operations, the institutional Army has in the main tended to regard small wars as distractions from the main task of preparing to fight great conventional engagements."[9]

Although the military certainly played a role in exploration, development, and expansion of the young nation,[10] the geopolitical uncertainties and interactions with the European great powers involved in the Napoleonic Wars kept military minds focused on traditional threats even while engaging with irregular ones. Both the War of 1812 and the Mexican-American War proved the importance of such a conventional emphasis, but the Civil War really cemented the prominence of traditional forms of war.

The stakes of the Civil War for the U.S. government (USG) and the scope and nature of the conflict ensured that, after initial attempts to put down the Confederate rebellion proved unsuccessful, the conflict would be a large-scale conventional fight.[11] The experiences of senior military leaders in the Mexican-American War and strategies and tactics limned from the Napoleonic Wars also reinforced the notion of big battle. The conduct of this war by the federal government would shape what Weigley termed the "American way of war." To him, the American military favored a strategy of annihilation focused on defeating and overthrowing the enemy. In contrast, a strategy of attrition based upon wearing down an enemy materially and temporally may have taken hold but lost out as American power and resources expanded. Later experience in World War II would reinforce these notions.[12]

Others, however, have disagreed with this interpretation. The military historian Brian Linn, for instance, has argued that "during the period between the end of the Civil War and the middle of the Second World War, the 'American way of war'—both in peacetime military thought and in wartime performance—was characterized more by improvisation and practicality than by a commitment to annihilation."[13] Max Boot, for his part, has asserted that small wars are the flip side of the coin of the traditional Weigley-esque American Way of War.[14] Similar to Boot, the international relations scholar Dominic Tierney delineates two American ways of war: interstate war and nation-building. In his analysis, while interstate wars are preferred both by the American people and by their military, and whereas American warfighting in the Civil War and World War II characterizes this "crusade tradition," the ambiguous, long-term COIN efforts from Southern Reconstruction to Vietnam to the present-day battlefields of Afghanistan make up a "quagmire tradition" of armed nation-building.[15] (Last, the soldier-scholar Jan K. Gleiman argues that there is an American counterculture of war of supporting insurgencies. That will be addressed in chapter 4.[16])

As will be shown later in this chapter, Linn, Boot, and Tierney offer many valid corrections or complementary addendums to Weigley's claims. For instance, between the Civil War and World War II, the U.S. Army fought not only in big battles in the Spanish-American War and World War I, but it also occupied the South after the Civil War, helped with expansion and fought against Native American populations in the West, and battled insurgents in the Philippines. The Marine Corps, for its part, also took part in World War I but carved out somewhat of a specialty as the so-called State Department troops engaged in murky small wars.

In fact, the Marine Corps developed such experience with small wars that they published the *Small Wars Manual* in 1940.[17] The Marines defined small wars as

> operations undertaken under executive authority, wherein military force is combined with diplomatic pressure in the internal or external affairs of another state whose government is unstable, inadequate, or unsatisfactory for the preservation of life and of such interests as are determined by the foreign policy of our Nation.[18]

The manual goes on to say that "small wars are conceived in uncertainty, are conducted often with precarious responsibility and doubtful authority, under indeterminate orders lacking specific instructions."[19] Topics ranging from COIN operations to the supervision of elections are covered within its pages. The necessity for the conduct of such operations, particularly within the U.S. geopolitical sphere of influence in the Western Hemisphere, made perfect sense as the nation rose from a commercial power to a great power.

The continued rise of American power after World War II and the releasing of the atom in 1945 would thereafter, until the late 1980s, cause American strategists to consider and think about fighting (thankfully) theoretical nuclear wars or real limited wars. The nuclear strategists, contemplating the most frightening dimension of warfare, were largely drawn from academia and examined the grandest dimensions of "grand strategy."[20] The limited war school's approach toward strategy, on the other hand, was that the application of force should be limited and should also be applied in appropriate levels so that its use would achieve diplomatic objectives.[21] Korea and Vietnam provided test cases for the application of limited war theory to the realities of combat.

Perhaps no event in the twentieth century affected the American military, foreign policy decision makers, and possibly the American public, as much as its involvement in Vietnam. Former chairman of the Joint Chiefs of Staff General Colin Powell, for instance, has stated:

> Our senior leadership knew the [Vietnam] war was going badly. Yet they bowed to groupthink pressure and kept up pretenses, the phony measure of body counts, the comforting illusion of secure hamlets, the inflated progress reports. As a corporate entity, the military failed to talk straight to its political leaders

> or itself. . . . Many of my generation . . . seasoned in that war, vowed that when our turn came to call the shots, we would not quietly acquiesce in halfhearted warfare for half-baked reasons that the American people could not understand or support.[22]

As General (GEN) Powell's quote expresses, that war has been *the* prime shaper of military beliefs for many of those officers who entered, from the late 1950s to the early 1970s, and remained within the services.[23] On the civilian side, talk of the "Vietnam syndrome" permeated debates over the use of force.

The aftermath of the Vietnam War, then, had several structural and institutional effects on the military. On the structural side, three factors seemed to be key determinants for examining the military. First, the United States ended conscription and then began the current All-Volunteer Force (AVF) in 1973. This move would have great long-term impact. Next, then-Army Chief of Staff General Creighton Abrams in the mid-1970s adroitly devised the "Total Force" concept. Under the Total Force concept, which still exists today, members of the National Guard and Army Reserve, the citizen-soldiers, would need to be mobilized in case of war and therefore, in theory, necessitate the public's support for large-scale overseas military operations. In other words, the U.S. Army would never have to go to war again without the support of the American people. (This was the theory at least. By the 1990s, select Guard and Reserve units were used more operationally and by the post-9/11 period the reserve had shifted fully from a strategic backfill to an operational force.) Last, the 1970s also saw a shift in the focus of the American military, and the army in particular, to the defense of Western Europe.

On the institutional side, the shift in focus for the American military to the defense of Western Europe proved valuable because it elicited a renaissance of strategic and military thought. A prime example of such thinking was the Clausewitzian revival embedded in the notion that our involvement in Vietnam necessitated the American military to require clear policy objectives and the ability to achieve them in order to commit itself to the use of force.[24] The election of Ronald Reagan as president in 1980 and his appointment of Caspar Weinberger as secretary of defense also had vast institutional impact.

In 1983, however, 241 U.S. Marines, sailors, and soldiers were killed by suicide bomber attacks in Lebanon while they were serving in a "peacekeeping" mission.[25] This event, and the supposed lessons of Vietnam, prompted Secretary Weinberger to deliver a speech at the National Press Club in 1984 in which he outlined six principles governing the use of American military force.[26] This speech dubbed, forever thereafter, the Weinberger Doctrine stated that American forces should not be used unless (1) vital national interests were at stake; (2) clear objectives for the use of force were stated; (3) the military was allowed to win; (4) congressional and public support existed for the intervention; (5) the intervention's purposes and objectives were monitored and adjusted; and, finally, (6) force should be used only as a last resort.[27] Several years later, this doctrine was refined by GEN Powell who added that U.S. forces should intervene decisively and use as much force as necessary in order to win wars quickly and to reduce the possibility of American casualties.[28] National security analysts have branded this style of warfare as "decisive force" and have noted accurately that this view's conception of the use of force often created tensions between the Pentagon and Foggy Bottom.[29]

The decisive force position was seemingly triumphant following the interventions in Panama in 1989 and the 1990–1991 Gulf War. The collapse of the Soviet Union and the geopolitical tremor and realignment of international power, however, ushered in a far less sanguine era than had been expected. Seemingly overnight preparations for fighting war against the Soviets on the Fulda Gap were replaced with the ambiguous and messy peacekeeping and peace enforcement operations as experienced in Somalia, Haiti, and the Balkans. Rather than referring to these operations as what they were, small wars, the Pentagon referred to them as military operations other than war (MOOTW). According to the then-official joint publication:

> **MOOTW focus on deterring war and promoting peace** while war encompasses large-scale, sustained combat operations to achieve national objectives or to protect national interests. MOOTW are more **sensitive to political considerations** and often the military may not be the primary player. More **restrictive rules of engagement** and a hierarchy of **national objectives** are followed.[30]

The vast scope and definitions of concepts such as MOOTW and "peace operations" made some observers nervous. Then-Army colonel Daniel Bolger, for instance, insisted that "when you carry a loaded rifle, get shot at, lose buddies, and live under the threats and stresses that go with those developments, it's war, plain and simple."[31] The frightening events that occurred in Mogadishu, Somalia on October 3–4, 1993, certainly lent credence to that position.[32] And throughout the 1990s as American strategic thinkers came to grapple with the ever so short "post–Cold War World," much debate surrounded the proper employment of the military abroad. Then-governor George W. Bush in the 2000 presidential election, after all, ran on a platform of keeping the United States out of nation-building.

Al-Qaeda's emergence in the 1990s appeared for many to be a minor irritant to a great power. Attacks on the East African embassies in Tanzania and Kenya and the attack on the USS *Cole* in the Aden harbor, while destructive, did not merit the serious attention that they deserved. The attacks against the World Trade Center and the Pentagon on September 11, 2001, would change that degree of focus. The strategic nature of the targets, the destruction and loss of life imposed, and the economic impact of the strikes seemingly changed the face of warfare.

After a successful unconventional campaign in Afghanistan and a similarly brilliant unconventional and conventional initial operation in Iraq, the U.S. military once again came face to face with the political-military complexities of small wars. COIN, which was largely neglected, wished away, and forgotten for a majority of the military since the Vietnam War, struck back. Resisted by both civilians at the Pentagon and by uniformed members of the military, learning and adaptation for COIN operations took hold.[33] The Army and Marine Corps Field Manual 3–24 *Counterinsurgency* amounted to a bureaucratic victory by a relatively small band of uniformed and civilian insurgents.[34]

The above having been said, there is still much debate about the future place of COIN as a mission area within the American military. After nearly two decades of war in Afghanistan, there are strong signs that elements of the civilian leadership, the military, and the American people are tired, or tiring, not only of the sacrifices in blood and treasure that the war has imposed but also in the seeming intractability of the conflict. Pro and con factions in the "COIN" debate have emerged.[35] Others meanwhile see such conflicts as the 2006 Lebanon War as the emergence of a hybrid form of warfare that operationally blurs irregular and traditional forms of warfare—and sometimes even adds the element of criminality.[36] Still others question whether large-scale conventional wars really are a thing of the past.[37] Former secretary of defense Robert M. Gates, for his part, attempted to allay fears by calling for a balanced approach to American military strategy[38] while his chairman of the Joint Chiefs of Staff Admiral Mike Mullen forcefully questioned whether the use of force should always be the last resort.[39]

When more discreet uses of force are called for, there seems to be an elite consensus that the "New Way of War" involves cyberwarfare and where terrorism will be dealt with by drones and SOF.[40] The Trump administration, meanwhile, at least when Secretary of Defense James Mattis was there, however, declared in the 2018 *National Defense Strategy*:

> Today, we are emerging from a period of strategic atrophy, aware that our competitive military advantage has been eroding. We are facing increased global disorder, characterized by decline in the long-standing rules-based international order—creating a security environment more complex and volatile than any we have experienced in recent memory. Inter-state strategic competition, not terrorism, is now the primary concern in U.S. national security.[41]

As stated in the prologue, that is a somewhat contentious argument that will be addressed in this book.

INSTANCES OF THE USE OF U.S. FORCES ABROAD, 1798–2018

But how does the aforementioned theory, historiography, and anecdotal evidence comport with the empirical record for the instances of the use of the U.S. armed forces abroad? Using numerous sources[42] a list of 314 such instances was compiled for the time period from 1798 to 2018.[43] These instances were subdivided between the "dominant mode of military activity" and their scale and duration axes. The dominant mode of military activity axis is subdivided between "irregular"[44] and "traditional"[45] operations and the scale and duration axis divides into four possibilities (small scale/short term, small scale/long term, large scale/short term, and large scale/long term). Table 1.1 graphically represents these divisions and includes all 314 instances of the use of U.S. forces abroad from 1798 to 2018 spread across the eight possible quadrant categories.[46]

What jumps out here is that for the U.S. military's so-called irregular operations are the dominant mode of military activity (with fully 73.4 percent of the cases) and that small-scale, short-term operations (of both the irregular and traditional varieties) are the historical norm (86.6 percent of the cases combined). Interestingly, while large-scale operations account for only 4.2 percent of the total number of cases, the figures for both irregular and traditional modes of military activity mirror one another. Last, aside from the small-scale/short-term cells, the largest remaining cell, accounting for 8.3 percent of the total number of cases, is for irregular small-scale/long-term uses of force.

But how do these cases look when they are disaggregated across different historical eras that account for varied levels of American international power? Using Max Boot's categories of eras of American power (commercial power, great power, and superpower[47]) yields the results found in tables 1.2–1.4.

Table 1.1. U.S. Uses of Armed Force Abroad, 1798–2018 (*N* = 314)

	Scale and Duration			
	Small Scale (1–39,999 Troops)		*Large Scale (40,000+ Troops)*	
Dominant Mode of Military Activity	Short Term *(< 2 years)*	Long Term *(> 2 years)*	Short Term *(< 2 years)*	Long Term *(> 2 years)*
"Irregular"	198 (63%)	26 (8.3%)	2 (0.6%)	5 (1.5%)
"Traditional"	74 (23.6%)	3 (0.9%)	2 (0.6%)	5(6)[1] (1.5%)

[1] The 6 in parentheses refers to the total number of such operations had the Civil War been counted as an operation abroad. A later chart includes the Civil War 1861–1865 era as a period of war in order to calculate "interwar periods" and so it is included here, but not counted in the total *N*.

Table 1.2. U.S. Uses of Armed Force Abroad in the Commercial Power Era, 1798–1898 (*N* = 95)

	Scale and Duration			
	Small Scale (1–39,999 Troops)		*Large Scale (40,000+ Troops)*	
Dominant Mode of Military Activity	Short Term *(< 2 years)*	Long Term *(> 2 years)*	Short Term *(< 2 years)*	Long Term *(> 2 years)*
"Irregular"	69 (72.7%)	4 (4.2%)	0 (0.0%)	0 (0.0%)
"Traditional"	19 (20.0%)	1 (1.0%)	0 (0.0%)	2(3) (2.1%)

Table 1.3. U.S. Uses of Armed Force Abroad in the Great Power Era, 1898–1941 (*N* = 64)

	Scale and Duration			
	Small Scale (1–39,999 Troops)		*Large Scale (40,000+ Troops)*	
Dominant Mode of Military Activity	Short Term *(< 2 years)*	Long Term *(> 2 years)*	Short Term *(< 2 years)*	Long Term *(> 2 years)*
"Irregular"	47 (73.4%)	9 (14.0%)	0 (0.0%)	1 (1.6%)
"Traditional"	5 (7.8%)	0 (0.0%)	1 (1.6%)	1 (1.6%)

Table 1.4. U.S. Uses of Armed Force Abroad in the Superpower Era, 1941–2018 (*N* = 156)

	Scale and Duration			
	Small Scale (1–39,999 Troops)		*Large Scale (40,000+ Troops)*	
Dominant Mode of Military Activity	Short Term *(< 2 years)*	Long Term *(> 2 years)*	Short Term *(< 2 years)*	Long Term *(> 2 years)*
"Irregular"	82 (52.6%)	13 (8.3%)	2 (1.3%)	4 (2.6%)
"Traditional"	50 (32%)	2 (1.3%)	1 (0.6%)	2 (1.3%)

As one might expect, the number of instances of the use of U.S. military forces abroad increases with the growth of American power. While there is less than one instance of the use of forces abroad per year during the Commercial Power era, those numbers jump to an average of 1.49 and 2 instances per year during the Great Power and Superpower eras, respectively. Irregular operations remain the dominant mode of military activity, accounting for

76.9 percent of the total number of instances in the Commercial Power era, a whopping 89 percent of the instances in the Great Power era, and a still robust 64.7 percent of instances in the Superpower era. In addition, and again unsurprisingly, the Superpower era accounts for 64 percent of the large-scale instances of the use of U.S. forces abroad.

Yet another way of disaggregating the use of forces abroad cases is to control for formal wartime periods.[48] Table 1.5 does this. It is interesting to note that the two most frenetic periods for the use of America's armed forces abroad were during the post–Spanish-American War to pre–World War I period and the post–Persian Gulf War to pre-9/11 interwar periods. Of course, the duration of the former interwar period covered over twice the amount of elapsed time as did the latter period. In other words, the U.S. Armed Forces were used abroad approximately once every six months during the post–Spanish-American War to pre–World War I interwar period as opposed to roughly once every eleven weeks from 1991 to 2001. Last, but not least, the interwar period between World War I and World War II was hardly a period of American "isolationism" militarily.

What can this empirical record tell us? Being fully aware of the warning contained in the fine print on investment brochures (i.e., past performance does not guarantee future returns), the assumption here based upon the analysis of the aggregate and disaggregated cases suggests that the United States will be engaged in various forms of small wars for the foreseeable future. As the strategic thinker and historian Eliot Cohen presciently wrote back in 1984 that if the United States were to reject forms of conflict then it will reject its status as a world power, in his words:

Table 1.5. Instances of the Use of Forces Abroad During Interwar Periods, 1783–2001[1] (*N*=243)

Interwar Periods	*Duration (in Years)*	*Instances of Use of Forces Abroad*	*Instance Ratio*
1783 to 1812	29	6	1 per 4.83 years
1815 to 1846	31	27	1 per 1.15 years
1848 to 1861	13	25	1 per 0.52 years
1865 to 1898	33	32	1 per 1.03 years
1898 to 1917	19	37	1 per 0.51 years
1918 to 1941	23	29	1 per 0.79 years
1945 to 1950	5	5	1 per year
1953 to 1964	11	9	1 per 1.22 years
1973 to 1990	17	30	1 per 0.57 years
1991 to 2001	10	43	1 per 0.23 years

[1] The end of the American Revolutionary War is added here for comprehensiveness.

> The choice, then, is between being ready to fight well when it is necessary to fight and fighting badly. The costs in the latter case include the loss of American and foreign lives and failures in American foreign and security policy. The best way to reduce the likelihood of small wars is to be well prepared for them, but it is a delusion to think that they are escapable through conventional deterrence, any more than they were in the late 1950s through a policy of massive nuclear retaliation. The first task, therefore, is to acknowledge the problem and study it.[49]

But along with such study, what is the best way in which to be engaged in them while still maintaining the ways and means to ensure national security?

AMERICAN POWER IN A DISORDERED WORLD

By 2020 the international system that the United States finds itself in is radically different than the one imagined a quarter century ago with the dissolution of the Soviet Union and the stunning military display witnessed in ejecting Saddam Hussein's army from Kuwait. Gone are the days of triumphalism about a unipolar moment.[50] Such unipolarity, if it ever existed at all, was, if anything, more a notion that U.S. conventional military capabilities lacked a peer competitor. But smaller, substate, and diffuse threats and challengers emerged almost immediately. As the late Harvey Sicherman wrote in 1997:

> The post–Cold War era may be distinguished by the absence of a cold war, but it hardly qualifies as "peace." Rather, an overarching geopolitical struggle has merely given way to numerous "underarching" struggles. The Kremlin may be emptied of ideologues plotting campaigns against America, her allies, and strategic points around the globe, but the world as a whole teems with as many geopoliticians as ever. One might call them the "meat-eaters," those to whom power, territorial possession, military action, and pirated wealth matter far more than do the environmental issues, diplomatic niceties, human rights, and other hobbies of the social transformers who comprise the current U.S. foreign policy establishment.[51]

For the first decade of the post–Cold War era, the vaunted U.S. military faced more irregular threats or nuisances operating in the political-military domain. But for all of the U.S. military dominance and discussions of a so-called revolution in military affairs,[52] gangs in Haiti, clans in Somalia, ethnosectarian militias in the Balkans, and Arab Afghan veterans of the Soviet-Afghan war and their spawn caused headaches for foreign governments and U.S. interests—and in some cases U.S. forces and embassies—in regions of the world that were sometimes seen as peripheral areas during the Cold War era.

State actors such as North Korea, Iran, and Iraq, referred to as rogue states, also pushed back against U.S. power.

But the attacks launched by al-Qaeda on September 11, 2001, in hindsight, began a more fundamental shift in the dynamics of the international system. For one, they punctured the illusion that the U.S. homeland was protected by the vast moats of the Atlantic and Pacific Oceans. For another they led to the invasions and overthrow of the Taliban government in Afghanistan and the Saddam Hussein regime in Iraq, which were stunning successes initially but were followed by long-term, messy COIN campaigns in both countries that proved especially difficult for U.S. forces and those of allies to cope with them, tied them down from responding to other contingencies, and the difficulties themselves may have emboldened certain actors within the system. U.S. forces were also deployed to the corners of the globe to assist friends against insurgents or instability in places such as Colombia, the Philippines, Georgia, the Pan Sahel and Maghreb regions of northern Africa, Yemen, Somalia, and elsewhere. At the same time, rising challengers such as China and Russia began to husband resources and dream of different distributions of international power. Such a quest for a recalibration of the pecking order of the international system was likely inevitable: after all, many in the 1990s saw a rising China as the next peer competitor to the United States,[53] but both the GWOT and the global financial crisis of 2008 and after perhaps accelerated this process.[54]

Thus, today the United States confronts a disordered world where China and Russia strain against the distribution of international power. Like all revisionist powers, they have for some time been conducting what might be called geopolitical probing.

China has probed by using its increased financial and military muscles by pressing its historical narrative about its claims over islands in the East China Sea and by literally constructing sovereignty through the building of islands and bases in the South China Sea and through the use of powerful soft power tools such as the development of parallel international institutions like the Asian Infrastructure Investment Bank that may seek either to supplement or supplant the post–World War II U.S.-led Bretton Woods system. Russia has similarly probed and rebelled against the post–Cold War status quo by inciting ethnic tensions in its "near abroad" and by its 2008 war with Georgia, its seizure of the Crimea in 2014, and its subsequent clumsily clandestine support for pro-Russian elements in Ukraine. It has carried out agitation and propaganda campaigns against the United States and the West. These gradual, iterative tactics track well with what Robert Strausz-Hupé and his coauthors described as the communist world's use of "nibbling expansion" against the Free World to avoid provoking a nuclear war with the United States.[55]

At the same time, the growth of the Islamic State in Iraq and Syria also exacerbated problems in the Greater Middle East. The Islamic State had used its particular vision of a strict form of Sunni Islam to win some degree of control over territory in Syria and Iraq. Their end-state was most revolutionary of all by seeking to overturn notions of man-made sovereignty altogether and to spread their self-proclaimed Caliphate to impose their stark interpretation of God's law upon all Muslims and to convert or kill those non-Muslims who will not. While they have not succeeded thus far, it is still a dangerous dream and appears to be an evergreen end-state for militant radical extremists. And this, of course, also plays out amid a broader geopolitical competition between the Sunni Gulf States and Iran with its tacit support, or sometimes imagined support, for Shi'a and proxy forces in Iraq, Syria, Lebanon, Yemen, and their co-religionists in the Gulf States. The Iranian quest for nuclear weapons has only increased concerns in both the Gulf States and in Israel.

On top of these shifts there still remain various state and substate security concerns relating to insurgencies, terrorism, proliferation, cyberattacks, and criminality—not to mention threats such as the spread of pandemics and climate change. The post–Cold War period of democratic transitions has also slowed, or in some cases backslid, with a resurgence in authoritarian or illiberal forces, whether religious, criminal, or ideological in nature, in a world where half of the sovereign nations are weak and are likely to remain so.[56]

The U.S. geographical location, economic strength (even if it is not what it once was in relative terms), and the global reach of its military power all mean that it will remain a, if not the, key player in international politics. No other state currently maintains the capabilities that the United States does. And certainly no liberal democratic state or block of democratic states currently can step in to provide the global capabilities and leadership that the United States can. For all the hopes that once blossomed about the possibilities of the emergence of a European state, those dreams are still a chimera and have been further challenged by the debt crises in places such as Greece. As the geopolitical analyst Robert Kaplan noted over a decade ago, there is no successor for American power for a more liberal world order. In his words:

> Winston Churchill saw in the United States a worthy successor to the British Empire, one that would carry on Britain's liberalizing mission. We cannot rest until something emerges that is just as estimable and concrete as what Churchill saw when he gazed across the Atlantic.[57]

This fact complicates matters.[58]

The hopes of the post–Cold War era led to an American grand strategy that might best be labeled as promoting "3P": peace, pluralism, and prosperity.

In other words, the foreign policy of the United States sought to promote international peace, promote or assist with the development of liberal democracies—or at least more democratic polities—abroad, and to expand domestic and international prosperities through the promotion of trade and by helping to maintain the global commons in the sea, air, and in the growing cyber domain. Presidents Bill Clinton, George W. Bush, and Barack Obama all have sought to follow this approach in the main, with each adjusting the ways and means to maintain the U.S. position in the world. President Donald J. Trump was less impressed with this approach.

President Trump's *National Security Strategy*[59] listed the U.S. vital interests as

- protecting the American people, the homeland, and the American way of life;
- promoting American prosperity;
- preserving peace through strength; and
- advancing American influence.

Rather than speak in more universalist language, it noted:

> Strengthening our sovereignty—the first duty of a government is to serve the interests of its own people—is a necessary condition for protecting these four national interests. And as we strengthen our sovereignty we will renew confidence in ourselves as a nation. We are proud of our history, optimistic about America's future, and confident of the positive example the United States offers to the world. We are also realistic and understand that the American way of life cannot be imposed upon others, nor is it the inevitable culmination of progress. Together with our allies, partners, and aspiring partners, the United States will pursue cooperation with reciprocity. Cooperation means sharing responsibilities and burdens. In trade, fair and reciprocal relationships benefit all with equal levels of market access and opportunities for economic growth. An America first National Security Strategy appreciates that America will catalyze conditions to unleash economic success for America and the world.[60]

But how does one place a hierarchy of interests? Derek S. Reveron and Nikolas K. Gvosdev offer a useful way of operationalizing such interests when they argue that vital interests are those that you are willing to die for, important interests are ones that you are willing to kill for, and peripheral interests are those that you are willing to fund.[61] Each administration places different values on what constitutes each of these types of interests—and particularly what makes up important and peripheral interests.

The United States, while fully capable of acting alone to project global military power, will prefer to work with allies and friends whenever possible when confronted with the necessity to threaten or use military force. This is the case not only to assist in burden sharing but also to increase international legitimacy for such operations—and this is particularly important when a United Nations Security Council (UNSC) Resolution cannot be obtained due to a veto by one of five permanent members. The North Atlantic Treaty Organization (NATO) allies are seen as prime partners to work with due to the existing and exercised coalition command and control arrangements, standard operating procedures, and logistical commonalities for things like munitions. The major non-NATO allies are also seen as key coalition partners.[62] Last, but certainly not the least, regional friends round out those that the United States prefers to work with, depending on the nature of contingencies.

As one might imagine, the relative importance of regions shift over time depending on the global geopolitical currents. Ten years ago, many would not include Europe as an important region of the world for the United States in the security realm, but Russia's revanchist behavior means that it is viewed as such today. The Greater Middle East, on the other hand, has seemed to remain a constant in terms of its importance since both the Soviet invasion of Afghanistan and the Iranian Revolution in 1979. The Persian Gulf War and the responses to 9/11 have kept it the most active theater of military operations for the United States. The world's dependence on oil and the vast production and energy reserves also are major factors for this emphasis.[63] Likewise, Asia's importance as a key hub in the global economy, particularly in terms of finance and manufacturing, has made it a key piece of geopolitical terrain. Meanwhile Latin America and Africa seem to ebb and flow in terms of the importance that policymakers in Washington place on them.

The tools of American power are multifaceted. The U.S. economy remains the largest in the world. Its legal system, with all of its blemishes and problems, is seen as a model of efficiency compared to much of the world. American culture, entertainment, and technological innovations are sought after around the globe. Its intelligence capabilities, particularly in the realm of collecting signals and electronic intelligence, are highly advanced. But most, including many Americans, focus on its military tool of power.

According to the International Institute for Strategic Studies' *The Military Balance 2019*, the U.S. defense budget in 2018 was $643.3 billion (which comprised a little more than 3 percent of the U.S. gross domestic product [GDP]), while the rest of the top fifteen nations spent roughly $734.5 billion (with China spending $168.2 billion), and the rest of the world spent a little more than $300 billion.[64] Many harp on these statistics to prove their point about an overly militarized United States. But, as stated above, the United States has global commitments and the key role in the maintenance of the global commons. The human and capital costs for maintaining the U.S. military are also much more expensive than for countries such as China and Russia. Added to this is the fact that many traditional U.S. allies have underinvested on defense, with many NATO members not even spending the specified alliance guidelines of investing 2 percent of member GDPs on defense.[65] President Trump has made a big deal of pressuring allies on burden sharing.

Still, while there are important reasons for large U.S. investments in defense, there has been somewhat of an imbalance in the mix of tools of American powers. For instance, the total U.S. budget is $4.41 trillion for the 2019 fiscal year (FY). Discretionary funding made up 31 percent of this figure, while 61 percent was nondiscretionary (Social Security, Medicaid, and Medicare), and 8 percent went to servicing the national debt. Of the discretionary funding portion of the budget, the military received over 50 percent of the pie. Total foreign economic aid, which includes the budgets of the State Department and the Agency for International Development (AID), accounted for $55.9 billion in FY20 and that number shrank under the Trump administration budget requests.[66]

This imbalance in spending between the military and the other elements of American power over a decade ago led the noted analyst Frank Hoffman to talk about the "one-armed cyclops" problem confronting American foreign and national security policy.[67] In this analogy, the cyclops's one muscular arm referred to military capability and the singular eye focused its attention on targeting threats. The scrawnier one, on Hoffman's telling, characterized the other tools of national power that were less developed, under resourced, and dulled compared to the Department of Defense's (DOD) tools. This has been a perennial issue in American foreign policy, particularly since the end of the Cold War. And it remains a topic of conversation largely because there is a degree of truth to it and, as the earlier numbers suggest, it is getting worse.

Without a doubt American conventional and strategic military power are the keys to its security from existential threats and are essential for its ability to help maintain the global commons and to project military power to the four corners of the Earth. And, to be sure, there are arguments to be made that the DOD may even be underfunded, given its level of global commitments (ranging from active combat operations to forward presence to humanitarian assistance). But, at the same time, the armed services cannot constantly look around for the "missing interagency" partners on the ground, but not also agree that those other partners do not require the necessary resources and capabilities. The zero-sum nature of resource constraints, at least under current conditions, leads to the proverbial "but that's my money!" problem in divvying up budgetary dollars.

At the same time, as the former deputy national security advisor Nadia Schadlow has argued, the nonmilitary and intelligence tools are less capable of thinking about preparing for and thinking about the current competitive era of world politics. In her words:

> Being successful in a competition requires knowing and understanding both one's competitors and oneself. Yet in those areas where non-military instruments of power dominate, the culture and the organizations needed to act competitively to achieve desired outcomes is generally absent. For the most part, competitive thinking is left to the realm of hard power. Only our military and intelligence agencies are structured to think and act competitively. The imbalance between military and non-military instruments of power is likely to continue unless civilian agencies develop approaches which account for the contested landscapes in which they function.
>
> A posture of competitive engagement would require that the civilian actors who oversee U.S. economic and humanitarian programs account for the fact that new ideas, economic strategies, civic action plans, and even public health-related initiatives are contested by vested interests or ideological or political opponents. This is true in a range of countries—from Pakistan, to Egypt, to Uzbekistan, to Somalia. It requires the recognition that even the building of a girl's school in Afghanistan or a health clinic in the Sudan is a political act. As the head of the Australian government's aid agency put it, "aid is 10 percent technical and 90 percent political."[68]

The strategic environment faced by the United States today is increasingly problematic for Americans in particular because we tend to view international politics and America's role in it as a dichotomy: we are either at war or at peace. But as the DOD's 2010 *Quadrennial Defense Review Report* pointed out, "The future strategic landscape will increasingly feature challenges in the ambiguous gray area that is neither fully war nor fully peace."[69] The Russian use of its "little green men" in Crimea and eastern Ukraine,[70] China's long-standing use of "slow intensity conflict" in the South China Sea,[71] and the complexities of disaggregating and working against various elements of combatants in Iraq and Syria (e.g., the forces of the Islamic State, the al-Qaeda-affiliated Nusrah Front, and various other groups) while trying to achieve our objectives, to include trying to find "the moderate opposition" we can work with and without getting so involved that we take ownership of the situation illustrate these problems.

As stated earlier, there will clearly always be a place for conventional military power. Today there should be an increasing debate about the role of unconventional military and paramilitary power. In the past, the unconventional element of such power was seen as a minor player that supplemented conventional military power by providing "economy of force"[72] elements or through the conduct of narrowly conducted special operations. But such either/or thinking about conventional and unconventional operations is as problematic as thinking about war and peace as pieces in separate, binary bins. The strength of the Russian, Chinese, Islamic State, and Iranian approaches in the current geostrategic environment is that they blend and blur military (conventional and unconventional) and nonmilitary approaches to press forward with their objectives.

As the United States continues to confront the challenges of the twenty-first century, its national security principals and agents should remember that strategy to the French General and strategic thinker André Beaufre

> is not like chess; its pieces have no permanent, defined value. It must therefore produce its solutions by a sort of cookery, fusing together constantly changing ingredients. War (or any contest) uses material force, the capabilities of which are determined by the material equipment available at the period. It also uses moral force; the latter is bound up with the ideas governing the civilization of the time. As a result strategy must be a continuous process of original thinking, based upon hypotheses which must be proved true or false as action proceeds; it is a process in which any mistakes of appreciation will reap the bitter reward of defeat. Herein lies strategy's great difficulty, especially in eras of rapid evolution like the present.[73]

This "cookery" is important across the domains of power.

As has been noted, today's highly competitive era of international politics spans all dimensions of state and nonstate power but mainly analysts focus on hard power tools such as military or economic considerations. But a downplayed tool of analysis is the role of competition in the political and informational domains—what can be labeled as "competitive shaping"[74]—that acts as a bridge and helps to synchronize and cleverly use and accentuate all elements of power whether one's level of analysis is global, regional, or national politics. Such competitive shaping ranges across a wide spectrum of things such as the attractiveness of a country's values or culture (to include its popular culture); passive or active (sometimes referred to as "political warfare") engagement with foreign labor movements, nongovernmental organizations, elements of civil society, and political movements; and usage of other elements of national power to promote strategic narratives (e.g., capabilities exercises for reassurance or deterrence).

Examples of such competitive shaping are Russia's promotion of the idea that peoples of Russian heritage, loosely defined, will be protected no matter what recognized international borders say and its use of such media outlets as RT to promote its agendas and expand its influence. (China's uses of such tools was discussed earlier.) Meanwhile nonstate actors such as the Islamic State and al-Qaeda seek to expand their power from the Maghreb to South and Central Asia and beyond through their own interpretations of Islam. Relying solely upon discrete diplomatic, economic, and military elements of power to try to come to grips with understanding these uses of competitive shaping or to try to develop strategies to counter them will offer incomplete answers no matter how well intended they might be.

These incomplete answers will be compounded by the fact that, as Nadia Schadlow has argued, many of the threats and challengers competing against the United States in the international system are much more adroit at thinking about, organizing for, and executing such competitive soft power than we are.[75] Furthermore, problems in dealing with non-Western states and cultures are compounded because, as the Naval Postgraduate School's Anna Simons contends, many of these states and cultures often know the United States and Americans better than we know them.[76]

CONCLUSIONS

This chapter set out to discuss the role of military power in U.S. foreign policy. While it showed and discussed how the military instrument is not the end-all-be-all tool of American power, it remains an often used tool. The historical record shows that the so-called irregular uses of force are actually the more regular application of the force of arms in the nation's history. While previous administrations and Trump's have sought to refrain from the use of force in such strategic environments that is usually easier said than done.

The remainder of this book will explore the more limited application of American military and paramilitary forces in dealing with these sorts of operational environments. The contemporary geostrategic environment seems to dictate that the United States must be able to operate more easily in the gray area, more aptly blend direct and indirect strategies to advance its interests and to counter those of its competitors, and more comfortably engage in competitive shaping. While the public's attitude toward involvement in further overseas adventures is less than sanguine and as budgetary follies will likely lead to reducing funding for defense and nondefense programs, it still remains important for the United States to be able to shape events on the ground overseas with as little force as possible or else live with the consequences of outcomes that may call for the use of more force down the road.[77]

NOTES

1. The late anthropologist Clifford Geertz defined culture as context in his classic work *The Interpretation of Cultures* (New York: Basic Books, 1977).

2. Carnes Lord very usefully defines strategic culture as "the product of very disparate influences. Foremost among them may be said to be the following: the

geopolitical setting in which a nation finds itself; its international relationships; its political culture and social structure; its military culture—military history, traditions, and education; its military and security organisations and their relationship to civilian authority; and weapons and technology. Strategic culture expresses itself in matters as diverse as strategic doctrine, personnel practices, command and control arrangements, and weapons procurement." "American Strategic Culture in Small Wars," *Small Wars and Insurgencies* 3, no. 3 (Winter 1992): 206–207. See also Basil Liddell Hart, *The British Way in Warfare* (London: Faber and Faber, 1932); Jack L. Snyder, *The Soviet Strategic Culture: Implications for Limited Nuclear Operations* (Santa Monica, CA: Rand Corporation, September 1977); Colin Gray, *Nuclear Strategy and National Style* (Lanham, MD: Madison Books, 1986); Alastair Iain Johnston, *Cultural Realism: Strategic Culture and Grand Strategy in Chinese History* (Princeton, NJ: Princeton University Press, 1995); and, Colin S. Gray, "Strategic Culture as Context: The First Generation of Theory Strikes Back," *Review of International Studies* 25, no. 1 (July 1999): 49–69.

3. Morton Halperin refers to such styles and preferences as organizational essence: "The organization's *essence* is the view held by the dominant group in the organization of what the missions and capabilities should be. Related to this are convictions about what kinds of people with what expertise, experience, and knowledge should be members of the organization." Morton H. Halperin with the assistance of Priscilla Clapp and Arnold Kanter, *Bureaucratic Politics & Foreign Policy* (Washington, DC: Brookings Institution, 1974), 28 (italics in original). Such organizational essences have also been referred to as service cultures. For examples of views about the service cultures of the U.S. military (which are sometimes lumped together as military culture), see Carl H. Builder, *The Masks of War: American Military Styles in Strategy and Analysis* (Baltimore, MD: Johns Hopkins University Press, 1989) and Thomas E. Ricks, *Making the Corps* (New York: Scribner Book Company, 1997).

4. The author's focus here is not to examine intrastate or state versus nonstate domestic uses of force in the United States because the scope of this book is to examine more recent uses of force abroad and therefore examining such domestic uses of force would confuse, more than clarify, comparative analysis.

5. Max Boot, *The Savage Wars of Peace: Small Wars and the Rise of American Power* (New York: Basic Books, 2002), xv–xvi.

6. Walter A. McDougall, *Promised Land, Crusader State: The American Encounter with the World since 1776* (New York: Houghton Mifflin, 1997). McDougall's view here is similar to Samuel P. Huntington's argument that "the American tends to be an extremist on the subject of war: he either embraces war wholeheartedly or rejects it completely. This extremism is required by the nature of the liberal ideology. Since liberalism deprecates the moral validity of the interests of the state in security, war must be either condemned as incompatible with liberal goals or justified as an ideological movement in support of those goals. American thought has not viewed war in the conservative-military sense as an instrument of national policy." Samuel P. Huntington, *The Soldier and the State: The Theory and Politics of Civil-Military Relations* (New York: Vintage Books, 1964; 1957), 151.

7. McDougall, *Promised Land, Crusader State*, 78. See also his *Freedom Just around the Corner: A New American History, 1585–1828* (New York: HarperCollins, 2005).

8. See, for instance, John Grenier, *The First Way of War: American War Making on the Frontier, 1607–1814* (New York: Cambridge University Press, 2008).

9. Colonel (ret.) John D. Waghelstein, "What's Wrong in Iraq? Or Ruminations of a Pachyderm," *Military Review* 86, no. 1 (January–February 2006): 112. See also his doctoral dissertation "Preparing for the Wrong War: The United States Army and Low-Intensity Conflict, 1755–1890" (PhD, Dissertation, Temple University, 1990). Waghelstein wrote his dissertation under the supervision of the late Professor Russell Weigley.

10. President Thomas Jefferson founded the U.S. Military Academy at West Point in March of 1802 in large part to produce engineers to develop the United States. See, for example, Robert M. S. McDonald, ed., *Thomas Jefferson's Military Academy: Founding West Point* (Charlottesville: University Press of Virginia, 2004). A year later, U.S. Army captain Meriwether Lewis selected William Clark to co-command his Corps of Discovery that explored what would eventually become the northern plains and northwestern United States. See, for instance, Stephen Ambrose, *Undaunted Courage: Meriwether Lewis, Thomas Jefferson, and the Opening of the American West* (New York: Simon & Schuster, 1997).

11. Just as in the case in the Revolutionary War, however, there were also irregular campaigns in the Civil War—particularly in the parts of the South and West. See, for example, Robert R. Mackey, *The Uncivil War: Irregular Warfare in the Upper South, 1861–1865* (Norman: University of Oklahoma Press, 2005); Thomas Goodrich, *Black Flag: Guerrilla Warfare on the Western Border, 1861–1865* (Bloomington: Indiana University Press, 1999); and Richard S. Brownlee, *Grey Ghosts of the Confederacy: Guerrilla Warfare in the West, 1861–1865* (Baton Rouge: Louisiana State University Press, 1984).

12. Russell F. Weigley, *The American Way of War: A History of American Military Strategy and Policy* (Bloomington: Indiana University Press, 1973).

13. Brian M. Linn, "'The American Way of War' *Revisited*," *The Journal of Military History* 66, no. 2 (April 2002): 503.

14. Boot argues that there are four varieties of such small wars: (1) punitive (punishing attackers of American citizens and property), (2) protective (safeguarding American citizens or property), (3) pacification (occupying foreign territory), and (4) profiteering (grabbing trade or territorial concessions). Boot, *The Savage Wars of Peace*, xv–xvi.

15. Dominic Tierney, *How We Fight: Crusades, Quagmires, and the American Way of War* (New York: Little, Brown and Company, 2010).

16. Jan K. Gleiman, "The American Counterculture of War: Supporting Foreign Insurgencies and the American Discourse of War," *Special Operations Journal* 1, no. 1 (2015): 19–36.

17. U.S. Marine Corps, *Small Wars Manual* (Washington, DC: Government Printing Office, 1940).

18. *Ibid.*, 1.

19. *Ibid.*, 9.

20. See, for example, Bernard Brodie, ed., *The Absolute Weapon: Atomic Power and World Order* (New York: Harcourt Brace, 1946) and Henry A. Kissinger, *Nuclear Weapons and Foreign Policy* (New York: Harper, 1956).

21. See, for example, Thomas C. Schelling, *Arms and Influence* (New Haven, CT: Yale University Press, 1966); and Morton H. Halperin, *Limited War in the Nuclear Age* (New York: Wiley, 1962).

22. Colin Powell with Joseph E. Persico, *My American Journey* (New York: Ballantine Books, 1996), 144–145.

23. For an outstanding account of how the war shaped the professional thinking of many respected high-ranking officers, see James Kitfield, *Prodigal Soldiers: How the Generation of Officers Born of Vietnam Revolutionized the American Style of War* (New York: Simon & Schuster, 1995).

24. The seminal work on this subject matter remains Col. Harry G. Summers Jr., *On Strategy: A Critical Analysis of the Vietnam War* (Novato, CA: Presidio Press, 1982). *Contra* Summers, however, Andrew F. Krepinevich, Jr. argues convincingly that the army's troubles in Vietnam did not occur due to a lack of focus on conventional tactics, but rather due to poorly conducted COIN. See his *The Army and Vietnam* (Baltimore, MD: Johns Hopkins University Press, 1988).

25. For a detailed account of U.S. Marine Corps involvement in Lebanon, see Daniel P. Bolger, *Savage Peace: Americans at War in the 1990s* (Novato, CA: Presidio Press, 1995), 167–219.

26. Caspar W. Weinberger, "The Uses of Military Power," *Department of Defense News Release* (November 28, 1984).

27. *Ibid.*

28. Powell with Persico, *My American Journey*, 420–421. This corollary to the Weinberger Doctrine has been termed the Powell Doctrine.

29. See, for example, F. G. Hoffman, *Decisive Force: The New American Way of War* (Westport, CT: Praeger Publishers, 1996); and Christopher M. Gacek, *The Logic of Force: The Dilemma of Limited War in American Foreign Policy* (New York: Columbia University Press, 1994).

30. *Joint Pub 3–07: Joint Doctrine for Military Operations Other Than War* (Washington, DC: Department of Defense, 16 June 1995), vii (bold in the original).

31. Bolger, *Savage Peace*, 38. Bolger is currently an army LTG.

32. See Mark Bowden, *Black Hawk Down: A Story of Modern War* (New York: Atlantic Monthly Press, 1999).

33. For more on learning and adaptation in COIN operations, see, for instance, John A. Nagl, *Learning to Eat Soup with a Knife: Counterinsurgency Lessons from Malaya and Vietnam* (Chicago: University of Chicago Press, 2005; Westport, CT: Praeger, 2002) and David H. Ucko, *The New Counterinsurgency Era: Transforming the U.S. Military for Modern Wars* (Washington, DC: Georgetown University Press, 2009).

34. Field Manual 3–24/Marine Corps Warfighting Publication 3–33.5, Counterinsurgency (Washington, DC: Department of the Army, December 2006). See also Conrad C. Crane, *Cassandra in Oz: Counterinsurgency and Future War* (Annapolis, MD: Naval Institute Press, 2016).

35. John Nagl and retired Army colonel Gian Gentile are seen by many as the leading proponent and dissenter, respectively, in the pro and con COIN schools. See, for instance, John A. Nagl, "Let's Win the Wars We're In," *Joint Force Quarterly* 52 (1st Quarter 2009): 20–26; and Gian P. Gentile, "Let's Build an Army to Win All Wars," *Joint Force Quarterly* 52 (1st Quarter 2009): 27–33. Both articles may be available online at: http://www.ndu.edu/inss/Press/jfq_pages/i52.htm (accessed May 1, 2012).

36. See, e.g., LTG James N. Mattis and Frank Hoffman, "Future Warfare: The Rise of Hybrid Warfare," *Naval Institute Proceedings* 131, no. 11 (November 2005): 30–32; Frank G. Hoffman, "How the Marines are Preparing for Hybrid Wars," *Armed Forces Journal* (April 2006). Available at: http://www.armedforcesjournal.com/2006/03/1813952/; Frank G. Hoffman "Preparing for Hybrid Wars," *Marine Corps Gazette* 91, no. 3 (March 2007): 57–61; Frank G. Hoffman, *Conflict in the 21st Century: The Rise of Hybrid Warfare* (Arlington, VA: Potomac Institute for Policy Studies, December 2007); Frank G. Hoffman, "Hybrid Threats and Challengers," *Joint Force Quarterly* 52 (January 2009): 34–39; William Nemeth, "Chechnya and the Future of Hybrid Warfare" (unpublished Master's Thesis, Naval Postgraduate School, Monterey, California, 2002); James Callard and Peter Faber, "An Emerging Synthesis for a New Way of War: Combinational Warfare and Future Innovation," *Georgetown Journal of International Affairs* 3, no. 1 (Winter/Spring 2002): 61–68; Michael Evans, "From Kadesh to Kandahar: Military Theory and the Future of War," *Naval War College Review* 61, no. 3 (Summer 2003): 132–150; Mackubin Thomas Owens, "America's 'Long War(s)'" FPRI E-Note, January 2008, available at: http://www.fpri.org/enotes/200801.owens.americaslongwars.html; John J. McCuen, "Hybrid Wars," *Military Review* 88, no. 2 (April–May 2008): 107–113.

37. See, for instance, Michael C. Horowitz and Dan Shalmon, "The Future of War and American Military Strategy," *Orbis* 53, no. 2 (Spring 2009): 300–318 and, Michael J. Mazarr, "The Folly of 'Asymmetric War'," *The Washington Quarterly* 31, no. 3 (Summer 2008): 33–53.

38. Robert M. Gates, "A Balanced Strategy," *Foreign Affairs* 88, no. 1 (January/February 2009): 28–40.

39. Admiral Mike Mullen, "Landon Lecture Series Remarks," Kansas State University, Manhattan, Kansas (March 3, 2010). Available at: http://www.jcs.mil/speech.aspx?id=1336 (accessed December 1, 2011).

40. See, for instance, Douglas A. Ollivant, "What War Has in Store," *CNN.com* (February 24, 2015). Available at: https://www.cnn.com/2015/02/23/opinion/ollivant-future-of-war/index.html.

41. *Summary of the 2018 National Defense Strategy: Sharpening the American Military's Competitive Edge* (Washington, DC: Department of Defense, 2018): 1. Available at: https://dod.defense.gov/Portals/1/Documents/pubs/2018-National-Defense-Strategy-Summary.pdf.

42. This information is derived from "Instances of Use of United States Armed Forces Abroad, 1798–2018," *CRS Report for Congress*, R42738, Updated December 28, 2018. A full listing of these operations is provided in this chapter's appendix A. The information for the appendix was gathered from this and the GlobalSecurity.org website (http://www.globalsecurity.org/military/ops/index.html); John M. Collins,

America's Small Wars: Lessons for the Future (Washington, DC: Brassey's [US], Inc., 1991); and author estimates. It is important to note that this author combined some of Grimmett's instances in order to attain simplification (e.g., actions involving the enforcement of the "no fly zone" over Iraq have been aggregated).

43. December 31, 2008, was selected as the end date in order to align this chapter's data set with Grimmett's report. See *ibid.*

44. "Irregular" operations consist of the threat or use of military force to accomplish policy objectives on the ground, at sea, or in the air as part of a whole of government approach in operational environments or situations where political, economic, or social considerations are in the lead in tactical operations.

45. "Traditional" modes of warfare place military considerations in the lead, driven by political purpose or pursuits.

46. While some may quibble with the simplification and coding of the cases, this is done for theoretical parsimony. As the late Samuel Huntington argued, "Understanding requires theory; theory requires abstraction; and abstraction requires the simplification and ordering of reality. No theory can explain all the facts. . . . Obviously, the real world is one of blends, irrationalities, and incongruities; actual personalities, institutions, and beliefs do not fit into neat logical categories. Yet neat logical categories are necessary if man is to think profitably about the real world in which he lives and to derive from it lessons for broader application and use. He is forced to generalize about phenomena which never quite operate according to the laws of human reason." Huntington, *The Soldier and the State*, vii.

47. Boot, *The Savage Wars of Peace*, xv–xvi.

48. Since the end of the Revolutionary War in 1783, the U.S. military has waged ten wars. Five of these conflicts were declared wars (the War of 1812, the Mexican War, the Spanish-American War, World War I, and World War II), one was the Civil War, and four were undeclared wars in the constitutional sense (Korea, Vietnam, the Persian Gulf War, and the GWOT/Iraq). This averages out to be roughly one war every 22.6 years.

49. Eliot A. Cohen, "Constraints on America's Conduct of Small Wars," *International Security* 9, no. 2 (Autumn 1984): 181.

50. See, for example, Charles Krauthammer, "The Unipolar Moment," *Foreign Affairs* 70, no. 1 (1990/1991): 23–33.

51. Harvey Sicherman, "The Revenge of Geopolitics," *Orbis* 41, no. 1 (Winter 1997): 7–8.

52. See, for example, Andrew F. Krepinevich, "Cavalry to Computer: The Pattern of Military Revolutions," *The National Interest* no. 37 (Fall 1994): 30–42.

53. See, for example, Richard Bernstein and Ross H. Munro, *The Coming Conflict with China* (New York: Vintage, 1998).

54. See, for instance, Aaron L. Friedberg, *A Contest for Supremacy: China, America, and the Struggle for Mastery in Asia* (New York: W. W. Norton & Company, 2011).

55. Robert Strausz-Hupé et al., *Protracted Conflict: A Challenging Study of Communist Strategy* (New York: Harper Colophon Books, 1963; 1959), xi.

56. Roy Godson, "Between Hard Power and Soft: We Need a New Type of Foreign Policy Professional," *The Weekly Standard* (February 6, 2012).

57. Robert D. Kaplan, "Supremacy by Stealth," *The Atlantic* (July/August 2003). Available at: http://www.theatlantic.com/magazine/archive/2003/07/supremacy-by-stealth/302760/.

58. See also Robert Kagan, *The Jungle Grows Back: America and Our Imperiled World* (New York: Knopf, 2018).

59. Available at: https://www.whitehouse.gov/wp-content/uploads/2017/12/NSS-Final-12-18-2017-0905.pdf.

60. *Ibid.*, 4.

61. Derek S. Reveron and Nikolas K. Gvosdev, "(Re)Discovering the National Interest: The Future of U.S. Foreign Policy and Defense Strategy," *Orbis* 59, no. 3 (Summer 2015): 306.

62. The major non-NATO allies are Afghanistan, Argentina, Australia, Bahrain, Brazil, Egypt, Israel, Japan, Jordan, Kuwait, Morocco, New Zealand, Pakistan, the Philippines, the Republic of Korea, Thailand, and Tunisia. Taiwan, due to the Taiwan Relations Act, will also be treated as if it were a designated a major non-NATO ally. See *22 CFR 120.32—Major Non-NATO Ally*. Available at https://www.law.cornell.edu/cfr/text/22/120.32.

63. Declining dependence on oil in the future, however, may change this emphasis.

64. International Institute for Strategic Studies, *The Military Balance 2019* (London: Routledge, February 2019), 21.

65. *Ibid.*, 59–60.

66. See www.usgovernmentspending.com. See also https://www.state.gov/wp-content/uploads/2020/02/FY-2021-CBJ-Final-508compliant.pdf.

67. Referenced in Michael P. Noonan, "The Future of American Military Strategy: A Conference Report," *FPRI E-Note* (February 2006). Available at: http://www.fpri.org/articles/2006/02/future-american-military-strategy-conference-report-0.

68. Nadia Schadlow, "Competitive Engagement: Upgrading America's Influence," *Orbis* (Fall 2013): 501–502.

69. "Quadrennial Defense Review Report" (Washington, DC: Department of Defense, February 2010): 73. Available at: http://www.defense.gov/qdr/images/QDR_as_of_12Feb10_1000.pdf.

70. See, for example, United States Army Special Operations Command, "Counter-Unconventional Warfare White Paper," Ft. Bragg, NC, 26 September 2014, pp. 29–32. Available at: https://info.publicintelligence.net/USASOC-CounterUnconventionalWarfare.pdf.

71. See, for example, Andrew Scobell, "Slow-Intensity Conflict in the South China Sea," *FPRI E-Note* (August 2000). Available at: http://www.fpri.org/articles/2000/08/slow-intensity-conflict-south-china-sea.

72. "Economy of Force: The Judicious Employment and Distribution of Forces so as to Expend the Minimum Essential Combat Power on Secondary Efforts in Order to Allocate the Maximum Possible Combat Power on Primary Efforts." *Joint Publication 1–02: Department of Defense Dictionary of Military and Associated Terms* (Washington, DC: Department of Defense, 8 November 2010—as Amended through 15 June 2015), 73.

73. General D'Armée André Beaufre, *An Introduction to Strategy: With Particular Reference to Problems of Defence, Politics, Economics, and Diplomacy in the Nuclear Age*, translated by Major General R. H. Barry (London: Faber and Faber, 1965), 136.

74. Adam Elkus and Michael P. Noonan, "Competitive Shaping in World Politics: A Bibliographic Essay and Course Outline," *The Philadelphia Papers* No. 18 (Philadelphia, PA: Foreign Policy Research Institute, May 2018). Available at: https://www.fpri.org/wp-content/uploads/2018/05/ElkusNoonanFINAL.pdf.

75. Nadia Schadlow, "The Problem with Hybrid Warfare," *War on the Rocks* (April 2, 2015). Available at: http://warontherocks.com/2015/04/the-problem-with-hybrid-warfare/.

76. Anna Simons, *21st Century Cultures of War: Advantage Them* (Philadelphia Paper No. 1, Foreign Policy Research Institute, April 2013). Available at: http://www.fpri.org/docs/Simons_21st_Century_Cultures_of_War.pdf.

77. Michael P. Noonan, "Re-Inventing Political Warfare," U.S. News & World Report's *World Report* blog (August 16, 2013). Available at: http://www.usnews.com/opinion/blogs/world-report/2013/08/16/political-warfare-in-a-time-of-defense-cuts.

Chapter 2

Indirect Strategy, Advisors, and the American Military Profession

In this chapter, the tenets of indirect strategy; the strengths and weaknesses of the use of military advisors working through, with, and by foreign forces; and the American military profession and its conceptions of advisors are discussed. These factors are important to consider to show the problems in developing the capabilities needed for fully implementing the defensive and offensive indirect interventions discussed in chapters 3 and 4, respectively. This is especially important to recognize and understand as these sorts of interventions are not going away.

As stated at the outset of this book, despite policy and the National Security Strategy, U.S. foreign policy and national security activity will likely still remain focused on conducting small-scale deployments and uses of force overseas and abroad. This will be the case because the national appetite for the large-scale deployment of troops will be low for the foreseeable future while interests, threats, and opportunities will call for the United States to remain engaged. Threats ranging from nonstate actors, such as the Islamic State, al-Qaeda, and international drug cartels, and state actors, ranging from Iranian Quds forces to Russian little green men to an increasingly globally assertive China, will ensure that economy of force considerations drive the United States to rely upon the use of small-scale military missions to act as force multipliers to promote with and through foreign partners and allies. The modern American way of war of cyber tools, drones, and direct action SOF raids will not always render efficient or effective results.

INDIRECT STRATEGY

The strategist Carnes Lord has argued that a large problem for Americans in the strategic realm is that they

> are a pragmatic people, tending to seek technical solutions to isolated problems and preoccupied with the here and now at the expense both of the past and the future. This means, among other things, that Americans tend to lack historical memory (critical for understanding other cultures) as well as the future orientation and holistic thinking that are the preconditions for strategy.[1]

If true, this insight is doubly problematic because the political-military orientation of all the so-called "small wars"[2] places a premium on one's ability to operate both within and among other cultures.[3] Compounding this perhaps is retired Army colonel and scholar Hy Rothstein's argument that "the U.S. military possesses a 'DNA' that can only produce one solution, a conventional solution at the attrition end of the attrition-maneuver continuum."[4] While some may consider this argument to be hyperbolic, it is not an uncommon argument for those who have operated regularly in such environs. This finding also argues against a body of scholarly literature that holds that democratic systems as a whole produce qualitatively superior military forces because of factors associated with regime-type, material preponderance, and other factors such as the initiative of their leaders or even culture.[5]

Operating in small-war environments requires much thought and consideration of nonmilitary factors by soldiers. This can lead to problems because, as the eminent soldier-scholar Sam C. Sarkesian has argued, nonmilitary factors are often seen as things that military commanders can simply add to operational orders rather than actually understanding and appreciating such things and how they can limit and restrain the broader environment.[6]

This is not to argue that conventional soldiers lack the mental acuity to think in nonmilitary terms, but it takes much honing.[7] The so-called Phase IV plans (i.e., what to do after the conclusion of conventional operations) in both Panama (1989) and Iraq (2003) are perfect examples of this phenomenon. Antulio Echevarria, a prominent military historian and strategist, has argued convincingly that this lack of holistic thinking about war and peace means that there is an "American way of battle" rather than a distinctive way of war.[8]

Today, with COIN fatigue almost completely settled in after nearly two decades of war and with the rise of India and China continuing to shift diplomatic, informational, military, and economic attention to East Asia and the reassertion of Russia in Eurasia, it may seem likely to some that the latest experiences with the messiness, ambiguity, and complexity of COIN will give way to a new era of international politics, or else return us to the former "comforts" of hard power in world affairs. Those that hold out hope for such a turn of events may be in for some disappointment. As mentioned earlier, the revisionist powers in the international system are engaging in geopolitical probing along and within the limits of the U.S. power commitments.[9] Such probes along the U.S. strategic perimeter may very well necessitate more small-scale advisory and training missions to support allies and friends or to create them.[10] Furthermore, the unsettled nature of the war in Syria and the remnants of both the Islamic State and al-Qaeda also complicates matters.[11]

For the French military theorist André Beaufre, strategy is "a *method of thought*, the object of which is to codify events, set them in order of priority and then choose the most effective course of action. . . . Any given strategy may be the best possible in certain situations and the worst conceivable in others."[12] Both direct and indirect strategies and admixtures of the two will therefore surely have their time and places. Strategy, however, is difficult because "the strategist is like a surgeon called upon to operate upon a sick person who is growing continuously and with extreme rapidity and of whose detailed anatomy he is not sure; his operating-table is in a state of perpetual motion and he must have ordered the instruments he is to use five years before-hand."[13]

The essential difference between indirect and direct strategies for Beaufre is the role of force. In indirect strategy, "the essential feature . . . is that it seeks to obtain a result by methods other than military victory."[14] Indirect strategy is very reliant upon carving out and making the best use of limited freedom of action. Whereas in direct strategy, "the essential element . . . is force, in other words material resources which, depending on their size, will enable us to obtain with greater or less ease the essential freedom of action."[15] Such indirect strategy is then very much in line with notions such as Lujan's "light footprints"[16] or Newson's "MINFORCE,"[17] but it still is not embraced by most segments of the American military profession.

Furthermore, operationalizing such indirect strategy troubles some. Brian Burton, for instance, wrote a cautionary article about the indirect approach in the National Defense University's journal *Prism*. There he argued that there were three perils to relying on such an approach: (1) getting the host nation (HN) to follow U.S. policy wishes and to comport with our norms for using force, (2) an inability to measure and assess the effectiveness of the HN's operations and built capacity, and (3) the difficulty of "off-ramping" from engagement once involved.[18] T. X. Hammes has also suggested that the very use of U.S. forces in expeditionary COIN on the ground will change the situation on the ground and therefore "the commander and his political bosses must be prepared to change the approach in kind."[19]

A more challenging problem to think through regarding the use of indirect approach is offered by the strategic analyst and author Carnes Lord when he stated:

> The political-military leadership, especially of a great power, rarely has time for sustained attention to a small war; yet such attention is often essential for crafting and implementing a national policy that fully integrates the diverse instruments at its disposal and applies them in a highly complex situation. Governments of great powers are especially vulnerable to what has been called the "political attrition" that can be inflicted by a weak yet resilient opponent. They are equally susceptible to manipulation by weak clients who engage them in a conflict to suit their own purposes.[20]

U.S. policymakers and their agents, for executing policy in the field and at American embassies, need to keep close tabs to ensure that allies or would-be allies are not inflating threats in order to gain access to American training and equipment that might be used for other purposes—particularly for maintaining internal security. Here, American provincialism may be of some assistance. As Hayward S. Florer, a retired Special Forces colonel, told Robert Kaplan, "Even our Special Ops people are insular. Sure, we like the adventure with other cultures, learning the history and language. But at heart many of us are farm boys who can't wait to get home. In this way we're not like the British and French. Our insularity protects us from becoming colonials."[21]

None of the above should be construed to argue that small footprint campaigns will be the future operational norm. One can, however, look at the historical trends in how the U.S. military has been used abroad and make the assumption that activities such as FID campaigns will be more likely than large-scale, long-duration COIN campaigns. But when they are called for, it will be imperative that the elements in the lead on the ground be the supported units.[22]

As the aforementioned Beaufre's cooking analogy argues, there will be numerous combinations for the use of the U.S. military abroad—acting either unilaterally or in concert with others. Table 2.1 offers a way to conceive of such forms of military involvement. And such involvement can take place from the so-called Phase 0 peacetime engagement all the way up through major combat operations and their aftermaths. Broadly speaking, U.S. forces can be used for offensive or defensive purposes—and sometimes both offensively and defensively simultaneously during conflicts such as World War II. Furthermore, such force employment might be thought of as indirect (low visibility) or direct (high visibility) and either limited (resource-constrained[23]) or assertive (less resource-constrained). Operations can transition between phases or categories (e.g., Operation Desert Shield was a defensive operation that transitioned to the offense with Desert Storm) and be ratcheted up or down according to changes in the strategic or political "ingredients."

Table 2.1. Forms of Military Involvement

	Offensive	*Defensive*
Indirect, Limited	Afghanistan (1979–1989), Iran (1980), Nicaragua (1981–1991)	El Salvador (1981–1992), Philippines (2001–2015), Trans-Sahara (2003–present)
Indirect, Assertive	Laos (1962–1975), Afghanistan (2001–2002), northern Iraq (2003)	Dhofar (1965–1975)
Direct, Limited	Somalia (1992–1993), Iraq (1998), Afghanistan and Sudan (1998), Kosovo (1999), Libya (2011)	Vietnam (1964–1965, 1972–1973), Afghanistan (2002–2008), Iraq (2009–2011)
Direct, Assertive	Grenada (1983), Panama (1989), Iraq (1991, 2003)	Vietnam (1965–1971), Saudi Arabia (1990) Afghanistan (2008–present), Iraq (2003–2009)

How might an indirect strategy be operationalized? Lieutenant Colonel Lujan suggests an eight-step checklist to the light-footprint approach:

1. **Prevention is the new "victory."** Preventing worst-case outcomes is the goal of light-footprint engagement and intervention, not decisive victory or transformation. Objectives should be modest, iterative, and consistent with the partner nation's (PN) historical and cultural contexts.
2. **Build and preserve networks.** Always build relationships, collect information, and develop understanding about potential security partners, even if conditions do not support providing formal U.S. military assistance. Engage well before a crisis and never completely disengage; recognize the potential future importance of networks.
3. **Partner with underlying interests and legitimacy in mind.** Before committing military advisers, training or equipment, ensure that the recipients' underlying interests, not just rhetoric, overlap with U.S. objectives and that the majority of the local population sees the partner as legitimate. Ongoing military support should be contingent upon demonstrated progress against mutually agreed-upon outcomes-based metrics, and the investment of U.S. resources should be targeted and small enough to make the threat of withdrawal credible.
4. **Less is more.** Engage with the smallest, lowest-profile military presence that can effectively foster critical indigenous security capabilities and protect U.S. interests.

5. **Minimize and phase out direct, unilateral action.** Grow indigenous capabilities and partnered relationships so that the potential need for direct U.S. military action becomes less likely over time. When offensive action must be taken, act indirectly, with local partners leading unless infiltration or other security concerns make this inadvisable. If a unilateral strike must be made, be as surgically precise as possible and balance the likely effects on the local political situation, the importance of the target, the imminent threat to the United States and the credibility of the information.
6. **Military engagement is a constructive tool, not a last resort.** Military engagement, partnership and training should be used proactively instead of as a last resort, and firmly integrated with a broader political, economic and diplomatic effort. Without a robust, civilian-led political plan, military efforts may only prolong the conflict.
7. **Do not surge for faster results.** Attempting to rush the pace of security force development and surge more resources than PN institutions can handle encourages corruption and decreases leverage because U.S. support will appear automatic and irreversible.
8. **Be prepared to fail.** Accept mission failure as a rare but normal outcome with abort criteria to support disengagement and prevent escalation in the event that the mission is no longer tenable or is not producing desired outcomes. Promote flat communications or direct feedback mechanisms between military and civilian leaders to continually manage risk, assess progress, and refine the mission via an iterative, two-way planning process.[24]

Another way to organize and implement such strategy might be through the so-called Global SOF Network. Scott Morrison argues that such a network would support U.S. defense strategy in three major ways by: (1) building trust and confidence across national SOF communities, (2) building up global capabilities for both self-defense and future coalition operations, and (3) providing U.S. SOF with "peacetime and contingency access."[25] (On this last point, such access is useful and necessary not only to build up partner SOF capacity, and our own, but also to be able to build up both regional-specific skills and also to be able to shape and react to emergent threats and opportunities.)

In 2013, U.S. SOCOM had opened an International Special Operations Forces Coordination Center. Countries represented included Jordan, South Korea, Italy, Poland, Thailand, Singapore, Japan, Lithuania, Hungary, and New Zealand, with another two dozen countries expected to be added within eighteen months.[26]

FOREIGN INTERNAL DEFENSE, UNCONVENTIONAL WARFARE, AND ADVISORS

The wars in Iraq and Afghanistan, as previously mentioned, raised American public consciousness, particularly for those too young to remember Vietnam, about counterinsurgency—sometimes shortened and referred to as "COIN." The DOD defines COIN as "comprehensive civilian and military efforts taken to defeat an insurgency and to address any core grievances."[27] The apparent success of the Iraq Surge beginning in 2007 and orchestrated by former secretary of defense Robert Gates, retired GEN David Petraeus, and others also contributed to the fairly broad awareness in certain circles of the Army and Marine Corps field manual on COIN.[28] But the manual itself, for some, seemed to be a one-size-fits-all approach to COIN where the large-scale employment of U.S. troops should always focus on a population-centric approach where they need to win over the support of the local populace in a battle of governance and legitimacy.[29]

Retired Marine Corps colonel T. X. Hammes has noted that population-centric COIN is just one of the possible permutations for the conduct of such operations. While he notes that some forms of COIN violate U.S. norms (such as models dependent upon such tactics as deportation, ruthless suppression, and in-migration), both decapitation (eliminating charismatic or very important insurgent leaders) and small, indirect approach COIN models may be more appropriate, depending on the circumstances on the ground.[30] He further cautions that the cost and low payoff for the United States in conducting population-centric COIN should steer decision makers away from these operations. The United States should only engage in such missions when it removes a foreign regime and must govern, but even then it must be "acutely aware of the time factor."[31]

Another voice cautioning against putting too much emphasis on large-scale, population-centric COIN is retired SF colonel David Maxwell. He notes that such a large-scale approach may be problematic because, "when US forces take the lead role using today's COIN theory and doctrine in actuality they are not conducting COIN since the insurgency is 'not theirs to counter' because the responsibility to counter it should belong to the sovereign nation that is faced with insurgency."[32] In other words, U.S. forces technically cannot be the counterinsurgents unless the insurgency is occurring in the United States.

Here we will focus on small-scale advisory efforts that generally fall into the category of "foreign internal defense." DOD defines FID as

> the participation by civilian agencies and military forces of a government or international organization in any of the programs or activities taken by a host

> nation (HN) government to free and protect its society from subversion, lawlessness, insurgency, violent extremism, terrorism, and other threats to its security. The United States Government (USG) applies FID programs or operations within a whole-of-government approach to enhance an HN internal defense and development (IDAD) program by specifically focusing on an anticipated, growing, or existing internal threat. A FID program would typically be supported by the Department of Defense (DOD) through routine security cooperation (SC) activities as part of the geographic combatant commander's (GCC's) theater campaign plan.[33]

FID is a core competency mission for SF.[34] This is unsurprising because, according to Sarkesian, "the very concept of Special Forces requires attention to political, social, and economic factors. The commitment of Special Forces must be in conjunction with a comprehensive strategy designed to support a whole range of political and economic programs in the indigenous system."[35] Others, however, have argued that post-9/11 operations have diverted SF away from such FID/UW capabilities and expertise and that renewed attention should be focused on this mission and mindset.[36]

Are SF best used as advisors? Is the indirect approach feasible and effective? In other words, rather than directly taking over or conducting a predominant portion of military operations as was the case early on in Iraq and still remains the case in Afghanistan, is the preferred model a small-scale one? According to Ernest Evans, the answer is yes because "a large-scale deployment of troops puts the prestige and credibility of the outside power on the line . . . a foreign country that overcommits itself in an insurgency greatly reduces its leverage with the host government because that government knows how difficult it will be for the foreign country to withdraw its troops."[37]

John Waghelstein, discussing a Special Forces Mobile Training Team (MTT) mission to support COIN operations in Bolivia, made a similar argument when he commented that "we did not take over the war, as sometimes tended to happen in those days, and our presence remained at a level that did not embarrass the Bolivians. The level of material support focused on what they really needed and not what they and the Embassy thought they needed."[38]

Echoing the earlier views is Ernest Lefever's observation that Military Assistance Program training is an extremely valuable tool to "encourage local and regional initiative in defense and security" and sustained assistance can lessen the burden on the United States and be lower cost, lower profile, and lower risk.[39] After all, as Stephen Peter Rosen has asserted, "Great powers need military allies, and when they cannot find them, they attempt to create them."[40]

As mentioned earlier, the experiences of Vietnam and the so-called Nixon Doctrine of Vietnamization and the shifting of burden sharing to allies in

Asia seemingly introduced the shift to FID.[41] But FID is not a one-size-fits-all approach. Joint Doctrine prescribes that three varieties of such operations exist:

- **Indirect support.** Indirect support focuses on building strong national infrastructures through economic and military capabilities that contribute to self-sufficiency, typically through SC activities.
- **Direct support (not involving U.S. combat operations).** These operations involve the use of U.S. forces to provide direct assistance to the HN civilian populace or military. Direct support operations are normally conducted when the HN has not attained self-sufficiency and is faced with social, economic, or military threats beyond its capability to handle.
- **U.S. combat operations.** The introduction of U.S. combat forces during FID requires a presidential decision and serves only as a temporary solution until HN forces are capable of conducting independent combat operations. Based on the assessment of the threat, the U.S. HN combat operations will likely take the form of one or more of COIN, CT, [counterdrug], or stabilization.[42]

This typology is useful as it is scalable and applies to a wide set of military activities, ranging from peacetime engagement to ambiguous gray zone deployments to the conduct of active armed hostilities.

In environments where insurgencies[43] exist, the external forces attempt to assist the HN government in holistically countering them. Some, such as retired Special Forces officer and former Pentagon official Kalev Sepp, argue that the best way to defeat such insurgencies is to drive a wedge between the population and the insurgents and to legitimize the HN government, "the focus of all civil and military plans and operations must be on the center of gravity in any conflict—the country's people and their belief in and support of the government."[44] Others, however, suggest that focusing on the enemy militarily is the best way to end such revolutionary action.[45] To be sure, the particularities of an insurgency—they are after all *sui generis*—will help to determine the best way, or ways, to counter it. Of course, such determinations are moot in a light-footprint advisory environment if advisors cannot assist their HN counterparts in coming to the right conclusions.

Military advice and assistance to HN governments generally come in three forms: (1) material advising and support (e.g., weapons and equipment), (2) nonmaterial advising and support (e.g., schooling and small unit training), or (3) a blended approach. The range and scope of such assistance are tied to the providing nation's foreign policy objectives and can take place in the provider's country, in the recipient's country, or in a third-party location. As Ernest Lefever argued, such military assistance training programs were to advance U.S. interests by promoting stability within and among participating states and their ability to defend themselves and to build person-to-person bonds and introduce them to American life and institutions.[46] One might term this the "capabilities and culture" approach.

In order for such a capabilities and culture approach to work, particularly in an irregular warfare combat zone, however, requires the right personnel. As Sam C. Sarkesian pointed out about U.S. advisors in Vietnam, many found their jobs frustrating and were unprepared to "cope with the myriad problems associated with an Asian military structure bound up in governing as well as conducting counter-revolutionary war."[47]

According to the political scientist and strategic analyst Stephen Peter Rosen, this push and pull of the political context creates a paradox: in order to create the fighting force the HN needs the advisor to become involved in local politics. His or her assets will be over the aid the government wants. In his words, "[their] biggest obstacle will be the fact that the host government will often place the creation of an effective army far below its own desire to stay in power."[48]

Following up on this point, the military anthropologist Anna Simons[49] has noted that "politics are inherent in all exchanges. Without question, too, the most effective military advisors have always recognized this and have used their status as non-natives to distinct advantage."[50]

Simons notes that advisors operate under the constraints that their security considerations are paramount, that they are initially always more dependent on their counterparts, and that they must accept that they are always being used by their counterparts. Most effective advisors learn to turn constraints into advantages because the advisory relationship is like other exchange relationships.[51] For instance, on the point about being used, she notes that advisors must understand this from the outset and "then employ *how* they are being used to their advantage."[52]

In addition, according to a recent analysis by retired lieutenant colonel Mark Grdovic, advisors must be able to build rapport with counterparts, have a mastery of tactical skills, and possess a macro-level understanding of their operating environment so that they can develop the micro-level understanding from their counterpart(s).[53] He goes on to say that ideally training should focus on those unique skills that are needed to serve as an advisor in a COIN environment. Language skills are imperative. Without those things advisors function as liaison officers, "who can provide some degree of ground truth about a situation but wield little or no real influence in shaping the environment in support of U.S. objectives." When that happens, historically, this has been the reason the military has resisted advisory efforts of marginal value and not worth significant investments.[54]

On the flipside of FID is UW. According to the DOD, UW consists of "operations and activities that are conducted to enable a resistance movement or insurgency to coerce, disrupt, or overthrow a government or occupying power by operating through or with an underground, auxiliary, and guerrilla force in a denied area."[55]

UW is the key mission of SF but other military and governmental agencies on the U.S. side and other friends and allies, third party and indigenous, may take part. According to retired Special Forces colonel Mark D. Boyatt, UW is political warfare "conducted *through, with and by* indigenous populations to achieve a legitimate political solution" and limits the exposure of a large number of U.S. military personnel.[56]

"Through" means to work *through* surrogates to conceal the U.S. role or when the U.S. military is not allowed direct contact with forces on the ground. "With" means to directly work *with* the party on the ground. Last, "By" means to work with the actors on the ground who carry out operations *by* themselves.[57]

UW will be addressed further in chapter 4.

FUTURE SECURITY COOPERATION AND ADVISING

When operating in small footprint environments, both SOF and general-purpose elements operate on the turf of both U.S. embassies and the regional COCOMs. The authorities granted to Chiefs of Missions (COMs) and the GCCs make coordination problematic, depending on one's interpretation of the statutes. Under U.S. Code Title 22, Section 3927, the duties of COMs and the duties of those assigned to their embassies are fully responsible for, under most circumstances, the "direction, coordination, and supervision of all Government executive branch employees in that country" and be apprised of all operations and activities of said employees in that country.[58]

Similarly, the duties and responsibilities of the commanders of the COCOMs[59] also fail to fully define "employees under the command of a United States area military commander." As such historical cases as El Salvador, Operation Enduring Freedom-Philippines (OEF-P), and Operation Enduring Freedom-Trans Sahara (OEF-TS) illustrate, ambiguities can cause problems the closer one moves down to the operational and tactical levels of analyses, particularly when the sticky variable of personalities comes into play. Without conclusive evidence of widespread, continuous problems with these seams of authorities, however, it would seem premature to make a definitive call for statutory clarification. But these issues do merit oversight scrutiny.

Turning our attention to the matter of advisory work over the past decade-plus of war has obviously spread the work around not only to members of U.S. SOF but also to members of conventional general purpose forces (GPFs) and even to a slew of private military contractors. This expansion of roles and missions has led to the creation of new terms and concepts.

Chief among these new terms and concepts is that of Security Force Assistance (SFA). SFA is defined by the U.S. Army as

> the unified action to generate, employ, and sustain local, host-nation or regional security forces in support of a legitimate authority (FM 3–07). Security force assistance (SFA) improves the capability and capacity of host-nation or regional security organization's security forces. These forces are collectively referred to as foreign security forces. ***Foreign security forces* are forces—including but not limited to military, paramilitary, police, and intelligence forces; border police, coast guard, and customs officials; and prison guards and correctional personnel—that provide security for a host nation and its relevant population or support a regional security organization's mission**.[60] [Boldface in original.]

Unlike FID, SFA may assist HN forces prepare for internal and external threats as well as be able to participate in international coalitions. It will be interesting to see if this mission area recedes in the future when the United States draws down its involvement in Afghanistan. Of course, with the inclusion of preparations to assist for dealing with external threats, SFA may be a growth area, particularly around and along the East Asian littoral.

In 2010, two SF majors with extensive GWOT experience, Jason A. Johnston and Stephen C. Taylor, wrote an important study about FID.[61] One limitation of contemporary FID that they saw "is that it is to be conducted when a [partner nation (PN)] faces internal threats only. Thus, as violent extremist organizations prove increasingly global in their reach and ambitions, they are outstripping what FID was designed to do. Arguably, this is the challenge we face in the [Federally Administered Tribal Areas (FATA)], which does not wholly belong to Pakistan."[62] In some respects, this ties in with SFA, but not quite.

Johnston and Taylor see three models for FID. One is highly visible U.S. training and advising with PN forces. Next is working in country, but not participating in tactical or combat operations. Last is the "indirect—train the trainer" technique of training and advising PN units in areas away from where they operate and "ideally is not even located in the PN."[63] In their opinion, the "indirect-train the trainer" approach is best. For them, "it is the only means that enables a nation to develop its own *Domestic Internal Defense* program, and, thereby, ensures a government is fulfilling *its* social contract with *its* citizens."[64]

Johnston and Taylor go on to state:

> Without question, there may be a need for unilateral action: this is called an invasion. There are also times for working alongside a foreign government's forces: this is called partnering. But then, perhaps more often than the U.S. government would like to believe, there are times and places when and where the United States should not participate in someone else's fight because our presence will only make things worse—for everyone. This describes the FATA right now.[65]

While the authors certainly make a compelling argument, this author is somewhat skeptical of the "indirect—train the trainer" model in a place like contemporary Pakistan. While such an approach certainly would make it easier in terms of domestic Pakistani political circumstances, the dangers of being too indirect seem to be that (1) it would be difficult to ensure that the Pakistani trainees would be using their acquired skills in an area such as the FATA instead of, say, Kashmir and (2) being too indirect would make it very difficult to monitor considerations such as the human rights practices of the HN forces or their commitment to civilian control of the military. While these quibbles may be seen as too focused upon the peculiarities of the case of Pakistan, upon reflection they seem to be useful considerations to think about when contemplating the use of an "indirect—train the trainer" model in other locales, too.

CONCLUSIONS

In terms of American military profession and civil-military relations theory, Samuel Huntington's 1957 fears[66] about subjective control and military fusionism were overstated and quite probably contributed to professional norms and biases that caused delays in the shifting of mental attitudes toward operational and tactical-level political-military issues among *some* general purpose forces (GPF) leaders—and even *some* SOF leadership—that suboptimized political-military performance in Afghanistan and Iraq. What will be interesting to see in several years is which military leaders climb to the senior ranks of the services after the United States departs Afghanistan and what forms will take the archetype "dominant military professionals."[67] Large-scale, long-term COIN campaigns will not likely be in vogue among senior leaders, not only because of the difficulties that they present but also because more offensive military doctrines tend to be favored by military organizations.[68] But, still, as Sam C. Sarkesian once pointed out:

> The wars of the future are likely to be an inexplicable mix of political, social, economic, and military factors. This will probably be true for all types of conflicts, ranging from nuclear war to limited military intervention. While the non-military issues surrounding any type of military action are an important part of all professionals' intellectual dimension, the aftermath of military action is even more important. In such circumstances, to presume that somehow one can clearly identify "pure" military issues and implement pure military solutions is simply to forget the lessons of history or to lack the intellectual dimensions for conceptualizing on such matters.[69]

One's enemies always get a vote in how they choose or will refuse battle and the way that they will counter their opponent across the spectrum of conflict.[70]

A more interesting question may be how will the experiences of the past decade affect civil-military relations among SOF elements—and in particular those that focus on FID and UW as discussed above? Eliot Cohen's argument about elite forces still resonates to a degree today when he argued that their availability and past success might distort perspectives on politico-military problems and seem to offer an easy way out of serious problems.[71]

But rather than applying to all of contemporary American SOF this warning probably applies most to the cadre of forces comprising the special mission units (SMUs) such as those that tracked down and killed Osama bin Laden in Abbottabad and other direct SOF[72] rather than the Special Forces, Civil Affairs (CA), and Military Information Support Operations forces as discussed earlier.[73] If anything, the latter forces are probably more susceptible to force structure cuts and budgetary cuts to the former SMU elements due to the future potential of deep defense spending reductions.[74]

SOF today, be they directly or indirectly focused, are not the French airborne forces of the Algerian war period. The fusionism skills[75] needed, particularly by indirect SOF, to ensure their political-military effectiveness are not calibrated for, nor desirous to, weakening or undermining civilian control in the United States. Rather, as Major Bradford Burris argued in the *Small Wars Journal*, SOF forces are "adhocracy" organizations whereas GPFs tend to be "professional bureaucracies."[76] These terms come from the writings of Henry Mintzberg. To Mintzberg, in adhocracy-based organizations, "nothing ever seems to get done without everyone talking to everyone else. Ambiguity abounds, giving rise to all sorts of conflicts and political pressures. Adhocracy can do no ordinary thing well. But it is extraordinary at innovation."[77] Professional bureaucracies, on the other hand, focus on the standardization of skills, operate best in complex-stable environments, and focus on the perfection of knowledge. This makes perfect sense. SOF select their members based on their drive, intelligence, physical fitness, and the ability to innovate. Advanced skills training and the psychological makeup of their members allow them to operate in a much more distributed way without the levels of support that GPF operate.

Still, as Anna Simons has noted, the conduct of advising and other operations in many parts of the globe have been complicated because foreign cultures have increasingly become much more aware about us than we are of them and that we do not specifically screen and recruit individuals for wile. In her words, wile:

> requires at least four attributes: 1) the ability to identify that feature or set of features in an adversary's culture that can be used as the fulcrum by, with, and through which to permanently alter conditions, 2) the ability to read all players so that you know how to appeal to, neutralize, and/or outwit each equally well,

> 3) an intuitive ability to tease, test, and probe so that you can make your own opportunities and don't operate on others' timeline(s), and 4) an appetite for twitting others, which means relishing the idea of turning the tables on adversaries in order to cause them to undo themselves.[78]

To maximize its effectiveness, when possessed, such wile must also be applied "a) locally, b) coordinating across locations and over time, and c) supralocally in order to both buy and control time."[79]

Developing such capabilities is important, because as the historian John R. Schindler has warned, for our adversaries and ourselves toward "special war," but this will be difficult for the United States because:

> special war is the default setting for countries that are unable or unwilling to fight major wars, but there are prerequisites, above all a degree of cunning and a willingness to accept operational risk to achieve strategic aims. . . . Special war works when competently handled. It's very cheap compared to any conventional military operations, and if executed properly it offers states a degree of plausible deniability while achieving state interests without fighting. The United States at present is not ready—organizationally, legally, politically, or culturally—to compete in special war. But getting proficient in special war will soon not be a choice, but a necessity.[80]

As we started this book discussing the differences between the use of raiding SOF versus more indirect approaches, one view offered by retired SF lieutenant colonel Tony Schwalm may be instructive. He compares the surgical strike-oriented SOF of the SMUs to Superman and the special warfare-oriented SOF of Special Forces to Daniel Boone. Describing the two approaches, he states that Superman "intends traumatic amputation" while Daniel Boone "brings cancer." Traumatic amputation kills faster, but UW is time-consuming. The SMUs devour money and equipment, and devour time, and these temporal realities (especially electorally) hurt the UW cause. In his words:

> Superman promises to get wherever the geopolitical fire is burning, put it out, and get back before anyone knows he's gone. Daniel promises to get to the fire, enter the house, find a corner not immediately threatened by conflagration, and begin an assessment process, the length of which may cost him his marriage. . . .
>
> Superman goes, does, and leaves. Daniel Boone goes, does, and stays and stays and stays. In the end, both come to the same place: killing somebody. The question then becomes who pulls the trigger.[81]

Politicians and military leaders prefer quick fixes. But those sorts of fixes are often quite unrealistic and ephemeral. The next two chapters deal with lengthier fixes, or attempted fixes, that are the hallmarks of such indirect interventions.

NOTES

1. Carnes Lord, "American Strategic Culture in Small Wars," *Small Wars and Insurgencies* 3, no. 3 (Winter 1992): 205–206. On the temporal dimension, the late Kevin Cunningham and Robert Tomes have similarly argued that Americans have a monochronic space-time orientation which causes them to view time as a measurable, zero-sum factor. Polychronic space-time orientations, on the other hand, view time as a nonzero sum, qualitative factor. See their "Space-Time Orientations and Contemporary Political-Military Thought," *Armed Forces & Society* 31, no. 1 (Fall 2004): 119–140.

2. Small-war environs here mean operational and tactical environments where political, social, and economic factors are just as much in the lead as military considerations.

3. This is not to argue that small wars are more complex than more traditional applications of force, but that with political considerations in the forefront of small wars operations the lack of cultural understanding makes the learning curve arguable much steeper among the human terrain. As the retired British GEN Rupert Smith argues, "In our modern conflicts, dealing with the civilian population is directly associated with the objective and is a primary, not secondary, activity. Furthermore, it is also the nexus for cooperation with all other agencies and levers of power in theatre. In practice such staff is in constant demand and in short supply, throwing a strain on the reserves and demonstrating their importance—yet they are frequently ill prepared for the task, drawn at random from other branches of the services, and on short tours in theatres that often demand the establishment of trust and good relations with the civilian population over time. Once again we have a prime example of the need to acknowledge the change in paradigm and adjust our militaries accordingly." See his *The Utility of Force: The Art of War in the Modern World* (New York: Vintage Books, 2007), 396–397.

4. Hy S. Rothstein, *Afghanistan & the Troubled Future of Unconventional Warfare* (Annapolis, MD: Naval Institute Press, 2006), 100. John Waghelstein has also stated that the DNA of the army rejects taking COIN seriously. See his quotes in Dexter Filkns, "The Fall of the Warrior King," *New York Times Magazine* (October 23, 2005). Available at: https://www.nytimes.com/2005/10/23/magazine/the-fall-of-the-warrior-king.html.

By the attrition end of the attrition-maneuver continuum, Rothstein is referring to Edward Luttwak's stated differences between forces designed for attrition vice relational-maneuver. These four "profound" differences are:

- Armed forces with a high-attrition content are supposed to optimize standard operating procedures for worldwide application, because for them all wars are the same. Low-intensity wars, however, are all different, and each requires an ad hoc set of standard operating procedures.
- Armed forces with a high-attrition content must treat all their personnel as interchangeable parts to maintain their efficiency. Low-intensity wars, on the other hand, usually require the persistent application of one-place, one-time expertise, embodied in specific individuals with unique attributes.

- Armed forces with a high-attrition content operate within an arena of military action demarcated by externally set political guides. Low-intensity wars, however, are made of political phenomena with a martial aspect. It follows that the senior officers of the dedicated body should have the particular aptitudes needed for the successful manipulation of the political variables. In low-intensity wars, victory is normally obtained by altering the political variables to the point where the enemy becomes ineffectual and not by actually defeating enemies in battle.
- Low-intensity wars cannot, by definition, be won by the efficient application of firepower. It follows that the officers of the dedicated body do not need the skills and aptitudes required for the management of large-scale organizations and the efficient operation of advanced equipment. On the other hand, they do need the ability to insert themselves into a foreign cultural milieu and to train and then lead local forces or native auxiliaries, who will almost always be equipped only with the simplest weapons.

"Notes on Low-Intensity Warfare," *Parameters* 13, no. 4 (Winter 1983): 16.

5. See, for example, Bruce Bueno de Mesquita and Rudolph M. Siverson, "War and the Survival of Political Leaders: A Comparative Study of Regime Types and Political Accountability," *American Political Science Review* 89, no. 4 (December 1995): 841–855; David A. Lake, "Powerful Pacifists: Democratic States and War," *American Political Science Review* 86, no. 1 (March 1992): 24–37; Dan Reiter and Allan C. Stam III, "Democracy and Battlefield Effectiveness," *Journal of Conflict Resolution* 42, no. 3 (June 1998): 259–277; and Victor Davis Hanson, *Carnage and Culture: Landmark Battles in the Rise of Western Power* (New York: Doubleday, 2001).

6. Sam C. Sarkesian, *Beyond the Battlefield: The New Military Professionalism* (New York: Pergamon Press, 1981), 188.

7. Once again, Sarkesian puts it best when he argued that "the intellectual perspective to understand and appreciate the inter-penetration between military and civilian elements, and the political and social consequences of military intervention cannot be achieved by theoretical studies on a one-time basis. This can best be acquired by a grasp of the tools of political-social analysis. Thus, from the time of planning, through execution and aftermath, the military professional will find it extremely difficult to limit himself solely to the military element. He must be able to assess the political-military and social inter-relationships, and their consequences on his own military efficiency. Such knowledge is developed over a period of time and requires serious study with intellectual tools that are rarely part of traditional military education and training." *Ibid.*, 172.

8. See his *Toward an American Way of War* (Carlisle, PA: Strategic Studies Institute, U.S. Army War College, March 1, 2004). Available at: http://www.strategicstudiesinstitute.army.mil/pubs/display.cfm?pubID=374 (accessed December 1, 2011). For a similarly convincing argument, see Nadia Schadlow and Richard Lacquement Jr., "Winning Wars, Not Just Battles: Expanding the Military Profession to Incorporate Stability Operations," in *American Civil-Military Relations: The Soldier and the State in a New Era*, edited by Suzanne C. Nielsen and Don M. Snider (Baltimore, MD: Johns Hopkins University Press, 2009), 112–132.

9. A. Wes Mitchell and Jakub Grygiel, "The Vulnerability of Peripheries," *The American Interest* 6, no. 4 (March/April 2011): 6.

10. Stephen Peter Rosen, "Brown Soldiers, White Officers," *The Washington Quarterly* 5, no. 2 (Spring 1982): 117.

11. Aside from current events in present-day Syria, the letters captured in the bin Laden compound in Abbottabad suggest such a shift in strategy. And events on the ground in both Yemen and Mali suggest such a shift with Ansar al-Shari`a and Ansar Dine working hand in hand with al-Qaeda in the Arabian Peninsula (AQAP) and al-Qaeda in the Islamic Maghreb (AQIM), respectively. See, for example, Nelly Lahoud et al., *Letters from Abbottabad: Bin Ladin Sidelined?* (West Point, NY: The Harmony Program, The Combating Terrorism Center at West Point, 3 May 2012), 48–52; and Christopher Swift, "AQAP, Ansar al-Shari`a and the Struggle for Yemen," *CTC Sentinel* 5, no. 6 (June 2012): 1–6.

12. General D'Armée André Beaufre, *An Introduction to Strategy: With Particular Reference to Problems of Defence, Politics, Economics, and Diplomacy in the Nuclear Age*, translated by Major General R. H. Barry (London: Faber and Faber, 1965), 13.

13. *Ibid.*, 46.

14. *Ibid.*, 108. The noncentrality of the role of military victory in indirect strategy tracks perfectly with the "quiet professional" ethos of the SF and on their shying away from taking credit in order to build up PNs and allies.

15. *Ibid.*, 122.

16. Major Fernando M. Lujan, *Light Footprints: The Future of American Military Intervention* (Washington, DC: Center for a New American Security, March 2013).

17. Rob Newson, "Adapting for the 'Other' War," *Small Wars Journal* (October 18, 2013). Available at: http://smallwarsjournal.com/jrnl/art/adapting-for-the-"other"-war.

18. Brian M. Burton, "The Promise and Peril of the Indirect Approach," *Prism* 3, no. 1 (December 2011): 58–59.

19. T. X. Hammes, "The Future of Counterinsurgency," *Orbis* 56, No. 4 (Fall 2012): 586.

20. Carnes Lord, "The Role of the United States in Small Wars," *Annals of the American Academy of Political and Social Science* 541, no. 1 (September 1995): 91.

21. Quoted in Robert D. Kaplan, "Supremacy by Stealth," *The Atlantic Monthly* 292, no. 1 (July/August 2003): 72.

22. David J. O'Hearn, Damon S. Robins, and Aaron C. Sessoms, "Flattening the Learning Curve: SOF as the Supported Command in the Irregular Warfare Environment" (Master's Thesis, Naval Postgraduate School, June 2012).

23. Such resource constraints may include limitations on personnel, equipment, aid, force capabilities (e.g., access to precision strikes or intelligence, surveillance, and reconnaissance assets), and the like.

24. Lujan, *Light Footprints*, 21–22.

25. Scott Morrison, "Redefining the Indirect Approach, Defining Special Operations Forces (SOF) Power, and the Global Networking of SOF," *Prism* 4, no. 3 (2013): 44–45.

26. Howard Altman, "Tampa to Become Epicenter of International Special Operations Coordination," *Tampa Tribune*, October 18, 2013.

27. "counterinsurgency," *DOD Dictionary of Military Terms*. Available at: http://www.dtic.mil/doctrine/dod_dictionary/ (accessed November 15, 2012).

28. Field Manual 3–24/Marine Corps Warfighting Publication 3–33.5, *Counterinsurgency* (Washington, DC: Department of the Army, December 2006). Available at: http://armypubs.army.mil/doctrine/DR_pubs/dr_a/pdf/fm3_24.pdf (accessed November 15, 2012). Such awareness was so broad, in fact, that the University of Chicago Press republished the Field Manual and one of the authors that helped to write it was invited to appear on Jon Stewart's popular *The Daily Show*. This is highly unusual for military doctrine. See, respectively, *The U.S. Army/Marine Corps Counterinsurgency Field Manual* (Chicago: University of Chicago Press, 2007) and "Lt. Col. John Nagl," *The Daily Show with Jon Stewart*. Available at: http://www.thedailyshow.com/watch/thu-august-23-2007/lt--col--john-nagl (accessed November 15, 2012).

29. See, for example, Frank G. Hoffman, "Neo-Classical Counterinsurgency?" *Parameters* 37, no. 2 (Summer 2007): 71–87.

30. Hammes, "The Future of Counterinsurgency," 565–587.

31. *Ibid.*, 583.

32. Colonel David Maxwell, "A Few Random Thoughts on COIN Theory and the Future," *Small Wars Journal* (September 26, 2009). Available at: http://smallwarsjournal.com/blog/a-few-random-thoughts-on-coin-theory-and-the-future (accessed November 15, 2012).

33. *Joint Publication 3–22: Foreign Internal Defense* (Washington, DC: Joint Chiefs of Staff, 17 August 2018), ix.

34. Other core SF missions include UW, direct action, and special reconnaissance. UW is the primary mission that U.S. SF are trained, organized, and equipped for and is defined as "activities conducted to enable a resistance movement or insurgency to coerce, disrupt, or overthrow a government or occupying power by operating through or with an underground, auxiliary, and guerrilla force in a denied area." FID in many ways offers the flipside of UW. See Field Manual 3–05, *Army Special Operations Forces* (Washington, DC: Department of the Army, September 2006). Available at: http://www.fas.org/irp/doddir/army/fm3-05.pdf (accessed November 15, 2012); and "unconventional warfare," *DOD Dictionary of Military Terms*. Available at: http://www.dtic.mil/doctrine/dod_dictionary/ (accessed November 15, 2012).

35. Sam C. Sarkesian, *The New Battlefield: The United States and Unconventional Conflicts* (New York: Greenwood Press, 1986), 233.

36. See, for example, Sean D. Naylor, "Support Grows for Standing Up an Unconventional Warfare Command," *Armed Forces Journal* (September 2007). Available at: http://www.armedforcesjournal.com/2007/09/3049653 (accessed June 1, 2012).

37. Ernest Evans, "El Salvador's Lessons for Future U.S. Interventions," *World Affairs* 160, no. 1 (Summer 1997): 46.

38. John D. Waghelstein, "Ruminations of a Pachyderm or What I Learned in the Counter-Insurgency Business," *Small Wars and Insurgencies* 86, no. 1 (Winter 1994): 362.

39. Ernest W. Lefever, "The Military Assistance Training Program," *Annals of the American Academy of Political and Social Science* 424, no. 1 (March 1976): 95.

40. Rosen, "Brown Soldiers, White Officers," 117.

41. On the Nixon Doctrine, see President Nixon's July 25, 1969, speech to journalists on Guam. Richard M. Nixon, *Public Papers of the Presidents of the United States, Richard Nixon: Containing the Public Messages, Speeches, and Statements of*

the President, 1969 to August 9, 1974 (Washington, DC: U.S. G.P.O., 1971–1975), 544–556. "FID as conducted today has its genesis in the post-Vietnam era when US policy shifted to emphasize that the United States would assist friendly nations, but would require them to provide the manpower and be ultimately responsible for their own national defense." *Joint Publication 3–22: Foreign Internal Defense*, ix.

42. Boldface in original. *Ibid.*, xi.

43. According to the French COIN theorist David Galula, such insurgencies are "protracted struggle[s] conducted methodically, step by step, in order to attain specific intermediate objectives leading finally to the overthrow of the existing order." See his *Counterinsurgency Warfare: Theory & Practice* (New York: Praeger, 1964).

44. Kalev I. Sepp, "Best Practices in Counterinsurgency," *Military Review* 85, no. 3 (May–June 2005): 9.

45. Nathan Leites and Charles Wolf, Jr., *Rebellion and Authority: An Analytic Essay on Insurgent Conflicts* (Chicago, IL: Markham, 1970) and Charles Wolf Jr., *Insurgency and Counterinsurgency: New Myths and Old Realities* (Santa Monica, CA: RAND, July 1965). Available at: http://www.rand.org/pubs/papers/P3132-1/ (accessed June 1, 2012).

46. Lefever, "The Military Assistance Training Program," 88. He later goes on to assert: "MAP training, by whatever name, is perhaps the single most valuable instrument the president has to encourage local and regional initiative in defense and security. A sustained training program leading to greater self-sufficiency and self-respect will have the effect of relieving the far-flung military burden of the United States. Military training is a low-cost and low-profile program. More important, it is a low-risk instrument" (p. 95).

47. Sarkesian, *Beyond the Battlefield*, 81–82.

48. Rosen, "Brown Soldiers, White Officers," 124.

49. Simons has spent much of her career examining military anthropology from a special operations forces perspective. For instance, see her *The Company They Keep: Life Inside the U.S. Army Special Forces* (New York: The Free Press, 1997).

50. Anna Simons, "The Military Advisor as Warrior-King, and Other 'Going Native' Temptations," in *Anthropology and the United States Military: Coming of Age in the Twenty-First Century*, edited by Pamela R. Frese and Margaret C. Harrell (New York: Palgrave Macmillan, 2003), 131.

51. *Ibid.*, 117–119.

52. Italics in original. *Ibid.*, 119.

53. Lieutenant Colonel Mark Grdovic, "The Advisory Challenge," *Special Warfare* 21, no. 1 (January–February 2008): 23–24.

54. *Ibid.*, 28. This advice rings true to the author who served as a mobilized U.S. Army Reservist officer on a military transition team (MiTT) with an Iraqi light-infantry battalion in 2006–2007. Overall, the training-provided pre-deployment was particularly underwhelming for the task at hand.

55. *Joint Publication 3–05 Special Operations* (Washington, DC: Joint Chiefs of Staff, 16 July 2014), xi.

56. Mark D. Boyatt, *Special Forces: A Unique National Asset "Through, With and By"* (Denver, CO: Outskirts Press, 2016), v.

57. *Ibid.*, 2–3.

58. "22 USC § 3927 - Chief of Mission." Available at: http://www.law.cornell.edu/uscode/text/22/3927 (accessed August 1, 2012). The statute states:

(a) **Duties**
Under the direction of the President, the chief of mission to a foreign country—
(1) shall have full responsibility for the direction, coordination, and supervision of all Government executive branch employees in that country (except for Voice of America correspondents on official assignment and employees under the command of a United States area military commander); and
(2) shall keep fully and currently informed with respect to all activities and operations of the Government within that country, and shall insure that all Government executive branch employees in that country (except for Voice of America correspondents on official assignment and employees under the command of a United States area military commander) comply fully with all applicable directives of the chief of mission.

(b) **Duties of agencies with employees in foreign countries**
Any executive branch agency having employees in a foreign country shall keep the chief of mission to that country fully and currently informed with respect to all activities and operations of its employees in that country, and shall insure that all of its employees in that country (except for Voice of America correspondents on official assignment and employees under the command of a United States area military commander) comply fully with all applicable directives of the chief of mission.

59. Appendix B has the relevant combatant commander authorities in full. It also offers the authorities and functions of U.S. Special Operations Command as a unified command.

60. Department of the Army, *FM 3–07.1 Security Force Assistance* (Washington, DC: 1 May 2009), 1–1.

61. Jason A. Johnston and Stephen C. Taylor, "Effective and Efficient Training and Advising in Pakistan" (Master's Thesis, Naval Postgraduate School, June 2010).

62. *Ibid.*, 29.

63. This technique, according to Johnston and Taylor "uses . . . indirect, low-visibility activities that support the legitimacy of the PN government by 'building strong national infrastructures through economic and military capabilities that contribute to self-sufficiency.' Indirect training and advising can include indirect support FID operations, such as JCETs and JCS exercises. Many indirect training and advisory activities fall under the U.S. Security Assistance (SA) Training Program, to include Foreign Military Sales, Professional Military Exchange, Unit Exchanges, and International Military Education and Training." *Ibid.*, 45.

64. *Ibid.*, 55–56.

65. *Ibid.*, 69.

66. And even in 1957, Huntington seemingly punted to a small degree on some political-military fusionism issues. For instance: "The military advisory groups to nations receiving American military assistance combined diplomatic and military responsibilities. By virtue of their control over the most significant form of American assistance, the advisory group and its chief at times tended to absorb duties and

influence which properly belonged to the ambassador and the State Department. While the problem of the organizing of civil and military representation in foreign countries on such a scale was relatively new to the United States, it had in the past been effectively dealt with by other nations. American military assistance and advisory groups will undoubtedly continue as a relatively permanent fixture in the conduct of American foreign affairs, and the eventual delimitation of responsibilities and duties between military and civil missions abroad should not present unsoluble difficulties." Samuel P. Huntington, *The Soldier and the State: The Theory and Politics of Civil-Military Relations* (New York: Vintage Books, 1964; 1957), 357.

67. Dominant military professional is taken from the works of Charles C. Moskos and John Allen Williams and is definitionally self-evident. Moskos argued that the post–Cold War era had ushered in a period where soldier-statesmen or soldier-scholars would supplant the former archetype of military leaders as managers or technicians. Williams extended this framework into the post-9/11 era and that myriad professional types would be emerging to include the soldier-warrior and manager, the soldier-statesman and scholar, and the soldier-constable. See, respectively, Charles C. Moskos, "Toward a Postmodern Military: The United States as Paradigm," in *The Postmodern Military: Armed Forces after the Cold War*, edited by Charles C. Moskos, John Allen Williams, and David R. Segal (New York: Oxford University Press, 1999), 14–31 and John Allen Williams, "The Military and Society Beyond the Postmodern Era," *Orbis* 52, no. 2 (Spring 2008): 199–216.

68. As Barry R. Posen has argued, militaries generally prefer offensive doctrines because they increase their autonomy vis-à-vis civilians. "Offensive doctrines tend to be more complicated than defensive or deterrent doctrines, and thus increase the difficulties for civilians who wish to understand military matters. Defense and deterrence are relatively easy for civilians to master." This is especially the case for nuclear and (counter)guerilla doctrines. See his *The Sources of Military Doctrine: France, Britain, and Germany Between the World Wars* (Ithaca, NY: Cornell University Press, 1984), 49–50.

69. Sarkesian, *Beyond the Battlefield*, 175.

70. As the French military officer GEN D'Armée André Beaufre pointed out: "If one side is at a considerable material disadvantage it can only hope to survive by refusing battle and using harassing tactics to keep the conflict going; this means guerrilla action which is as old as the hills but which each generation has nevertheless forgotten and had to learn afresh." See Beaufre, *An Introduction to Strategy*, 114.

71. Eliot A. Cohen, *Commandos and Politicians: Elite Military Units in Modern Democracies* (Cambridge, MA: Center for International Affairs, Harvard University, 1978), 78–79.

72. By direct SOF, the author means units such as non-SMU Navy SEAL units, the Marine Special Operations Command forces other than their foreign military training units, elements of the Air Force's Special Tactics Squadrons, and Army Rangers.

73. For these units, the author uses the shorthand of indirect action SOF. This is strictly for categorical parsimony and grouping.

74. Defense cuts will also cause issues in terms of maintaining both SMU, non-SMU direct-action-focused SOF, and indirect-action-focused SOF because all of their

members are drawn from the services' GPFs or are recruited by the GPFs. As Morton H. Halperin argued back during the Vietnam era: "Career Army officers agree that the essence of the Army is ground combat capability. They tend to deprive of funds those functions which they view as peripheral, such as advisory roles in Military Assistance Advisory Group (MAAG) missions, air defense, and the so-called 'Green Beret' counterinsurgency forces." See his, with the assistance of Priscilla Clapp and Arnold Kanter, *Bureaucratic Politics & Foreign Policy* (Washington, DC: Brookings Institution, 1974), 34. What will be interesting about future cuts will be to see what happens between the direct and indirect SOF elements and their allocations of the respective budgetary pie. Of course, the existence of U.S. Special Operations Command today as well as supporters in the White House and on Capitol Hill might save SOF from deep funding cuts.

75. Indirect SOF have taken to heart Sarkesian's warning that "the intellectual perspective to understand and appreciate the inter-penetration between military and civilian elements, and the political and social consequences of military intervention cannot be achieved by theoretical studies on a one-time basis. This can best be acquired by a grasp of the tools of political-social analysis. Thus, from the time of planning, through execution and aftermath, the military professional will find it extremely difficult to limit himself solely to the military element. He must be able to assess the political-military and social inter-relationships, and their consequences on his own military efficiency. Such knowledge is developed over a period of time and requires serious study with intellectual tools that are rarely part of traditional military education and training." Sarkesian, *Beyond the Battlefield*, 172.

76. Bradford Burris, "Professional Military Education for United States Army Special Operations Forces," *Small Wars Journal* (August 19, 2010). Available at: http://smallwarsjournal.com/jrnl/art/pme-for-us-army-special-operations-forces (accessed on August 1, 2012).

77. Henry Mintzberg, "Organizational Design: Fashion or Fit?" *Harvard Business Review* 59, no. 1 (January–February 1981): 12.

78. Anna Simons, *21st Century Cultures of War: Advantage Them* (Philadelphia Paper No. 1, Foreign Policy Research Institute, April 2013), 38.

79. *Ibid.*

80. John R. Schindler, "We're Entering the Age of 'Special War'," *Business Insider* (September 25, 2013). Available at: https://www.businessinsider.com/were-entering-the-age-of-special-war-2013-9.

81. Tony Schwalm, *The Guerilla Factory: The Making of Special Forces Officers, The Green Berets* (New York: The Free Press, 2012), 5.

Chapter 3

Defensive Indirect Interventions—Foreign Internal Defense

This chapter continues the discussion of FID from chapter 2 and then uses the examples of El Salvador (1980–1992), OEF-P (2001–2015), and OEF-TS (2002–2015) to illustrate and illuminate limited defensive indirect interventions. The British case of Dhofar, Oman (1965–1976), is then used to examine a limited assertive indirect intervention because a suitable U.S. case did not exist. These cases offer useful insights on the possibilities and pitfalls of such defensive indirect interventions. Such examples are helpful because this is an area where the United States will still be engaged as it confronts both GPC and other state and nonstate actors.

Even a nation with robust military capabilities and the ability to globally project its power cannot be in all places at all times. Countries judge, weigh, and adjust their security commitments based upon myriad considerations. When concerns about regional security issues arise, a country may, for instance, do nothing, monitor the situation, work through allies, or else make commitments across the elements of its national power when they seek to assist a government in defending its sovereignty. Furthermore, such commitments might take the form of the provision of development aid, intelligence support, equipment, or the introduction of military or paramilitary forces. The involvement of such forces, as chapter 1 illustrated, might be large scale (e.g., Vietnam, Iraq, and Afghanistan) or they may be a handful of individuals (i.e., Colonel Edward Lansdale and the Philippines in the 1950s).[1] This chapter deals with the lower end of this spectrum when a nation goes to support an HN government to defend itself from internal threats—or internal threats with external support—without the large-scale deployment of troops or other capabilities.

INDIRECT INTERVENTIONS AND FOREIGN INTERNAL DEFENSE

As was pointed out in the previous chapter, FID in U.S. doctrine involves the "participation by civilian agencies and military forces of a government or international organization" to assist an HN government to "free and protect its society from subversion, lawlessness, insurgency, violent extremism, terrorism, and other threats to its security."[2] It is important to note that this definition makes mention of nonmilitary actors. An HN facing internal security threats from, say, narcotic traffickers may want the assistance of the U.S. Drug Enforcement Administration, for instance. (In fact, just this sort of civilian agency assistance to FID can be seen in the Netflix programs *Narcos* and *Narcos: Mexico*.[3])

For analytical simplification, small footprint FID here is split between the categories of limited (i.e., the use of U.S. forces in combat operations is restricted) or assertive (i.e., U.S. combat involvement is relatively unrestricted). Using the "through, with, and by" construct discussed previously, all three might take place in a limited defensive intervention, whereas an assertive defensive intervention would entail a robust form of "with." Under present doctrine, this would be referred to "US combat operations."[4] Table 3.1 offers some examples of each category.

Whether the United States, or another country or bloc of countries, intervenes in a limited or assertive fashion, the desired end-state is the same: to allow the HN government to develop its own forces and capabilities to deal with its own security needs. In this way, any such intervention follows under what might be termed the "BASF Doctrine": we don't fight your insurgency, lawlessness, terrorism, or criminality. We help you fight them better.[5] Unfortunately, however, as Colonel Brian Petit has asserted, such forms of intervention often do not lead to quick, measurable results. "Measures of effectiveness are often best assessed over time and anecdotally,"[6] he notes.

The external intervener must not only work with the HN on the ground comprising a multitude of actors (e.g., the government, police, military, disparate communities, and nongovernmental agencies) but also work among its own governmental agencies and possibly with those of other outside forces. Differing bureaucratic or personal agendas and favored approaches and the

Table 3.1. Forms of Defensive Indirect Interventions

Limited	For example, El Salvador (1981–1992), Plan Colombia (1999–2015), Philippines (2001–2015), Georgia Train and Equip (2002–2004), Pan Sahel/Trans-Sahara (2003–present), Central Africa (2011–2017)
Assertive	For example, Dhofar (1965–1975) [non-U.S. case], Iraq (2014–present)

possibility of personality conflicts can complicate assistance. Such conflicts also may mean that common agreement on what constitutes effectiveness or success may be difficult to define.

But just as a nation like the United States may intervene to bolster the capabilities of the HN government, other actors, be they states or nonstate forces, may also intervene to help in fostering subversion, lawlessness, insurgency, violent extremism, terrorism, or criminality. Such interventions may be overt, covert (i.e., plausibly deniable), or clandestine (i.e., tactically secret) in nature.[7] The U.S. Army Special Operations Command published a white paper on this entitled *Counter-Unconventional Warfare*.[8] Such counter-UW operations seek to "decrease the sponsor's capacity to employ unconventional warfare to achieve strategic aims."[9]

LIMITED DEFENSIVE INDIRECT INTERVENTIONS

Three small footprint-limited FID missions are discussed: El Salvador (1981–1992), the Philippines (2001–present), and the Pan Sahel/Trans-Sahara region (2002–present). Why focus on these three cases? El Salvador is quite obvious as it has been pointed to as an example of a successful small-scale advisory campaign. It is also useful because of its Cold War context. In an era of GPC, it is beneficial to examine a case that took place as part of a larger geopolitical contest. The Philippines and Pan Sahel/Trans-Sahara cases were selected because they show such engagement in one country and across a region. In addition, the extended analyses of these cases are intended to ensure that their identifiable lessons are not lost in all of the continuing focus on Afghanistan and Iraq and they are more appropriate harbingers of the future than those two large footprint/combat operations heavy wars.[10]

El Salvador, 1981–1992

The U.S. involvement in the El Salvadoran civil war has to be recognized within its Cold War context. U.S.-Soviet competition led to confrontations between proxies across the globe. Since the Cuban Revolution, there was an expansion of support to communist insurgents through Latin America. El Salvador proved quite susceptible to such communist agitation due to several underlying factors.

El Salvador is a country slightly smaller than the state of Massachusetts and located toward the northern end of Central America. The only nation in Central America without a coastline on the Caribbean Sea, it is situated south of Guatemala (sharing a 126-mile land border), west of Honduras (212-mile border), northwest of Nicaragua across the Gulf of Fonseca, and has a 190-mile coastline on the Pacific Ocean.[11] The country's topography is quite rugged and contains several dormant volcanoes and has a history of seismic activity. In 1980, it possessed an extremely high population density of approximately 230 people per square kilometer. The estimated population grew from over 3.5 million people in 1971 to nearly 5.4 million people in 1988. Coupled with this was the fact that the coffee was responsible for 80 percent of GDP and that 60 percent of the population owned no lands and were essentially seasonal migrant workers on the coffee plantations. About 4 percent of the population owned about 65 percent of the lands.

Politically, the country achieved independence from Spain in 1821 along with the other states of Central America and left their United Provinces government for total independence in January of 1841. Economically, first indigo and later coffee came to dominate. This led eventually to the economic centralization of power among a set of oligarchic families (*La Catorce*) and later politically with strongman rulers. A sharp economic downturn between 1928 and 1931 caused local peasants (*campesinos*) to rally around Marxist-inspired radicals such as Agustin Farabundo Marti. In 1931, the military launched a coup d'état to crack down on leftist political activities. When Marti and others launched an uprising in 1932, it was brutally put down. Estimates suggest that up to 30,000 *campesinos* were killed in what came to be known as *la matanza* and fueled lingering resentments that would last for decades.

The 1969 "Football War" with Honduras exacerbated economic (lost trade) and demographic (post-war Honduran crackdowns on illegal emigration removed a safety valve) problems for El Salvador. Further political difficulties led to a coup in 1972 against President Jose Napoleon Duarte, who was forced into exile. This led to further instability and agitation in the country. After a rigged election brought GEN Carlos Humberto Romero Mena to the presidency in 1977, a series of three reformist coups took place after 1979 that led the country to a path of open revolt.[12]

After coming into office decrying an "inordinate fear of communism,"[13] the presidential administration of Jimmy Carter started to take notice of the spread of Soviet-backed revolutionary regimes in Latin America, Africa, and Asia. With the fall of nearby Nicaragua to the Soviet- and Cuban-backed Sandinista regime, the Carter administration restarted military aid to El Salvador in the spring of 1980 to confront the Farabundo Marti National Liberation Front (FMLN) insurgency named after the rebel martyr from the 1930s. This would be the beginning of a decade-plus-long American involvement there.

Lieutenant General (LTG) Wallace H. Nutting, U.S. Army, arrived as the Commander-in-Chief (CINC),[14] U.S. Southern Command (SOUTHCOM), in 1979, fresh from the front lines of the conventional Cold War on the Fulda Gap in Germany. He felt that his career had been prematurely cut short. The two CINCs adjacent to his command—Pacific Command (PACOM) and the then-Atlantic Command, respectively—outranked him as full admirals and he had only ninety-four personnel assigned to his headquarters when he arrived. Lacking strategic guidance from Washington, he set about devising his own theater strategy and moved away from the focus on the Panama Canal toward supporting a COIN campaign in El Salvador.[15]

According to Kalev Sepp, three themes shaped LTG Nutting's environment: (1) a lack of intergovernmental coordination (most strikingly, as stated earlier, no policy to guide strategy), (2) intramilitary reluctance and resistance (especially relating to the distribution of resources; SOUTHCOM's portion of the SA budget was less than 2 percent of the total), and (3) the effect of the defeat in Vietnam on American political and military culture.[16] On top of this, the U.S. ambassador to El Salvador, Robert E. White, was also cool toward the general and would not allow him to travel into the country.[17]

One asset that LTG Nutting did have assigned to him in Panama was the Third Battalion of the 7th Special Forces Group (Airborne). The 3/7SFG(A) was a leftover remnant from the 8th Special Forces Group, which was deactivated in the Panama Canal Zone in 1972.[18] Coinciding with a large demonstration that was taking place in El Salvador on January 23, 1980, SOUTHCOM directed the Commander of 3/7SFG(A) to prepare a large-scale, long-term advisory effort for El Salvador. The battalion staff prepared the proposal in a single day:

> The phased program called for an initial 55-man training and assistance group, to coordinate Salvadoran intelligence, operations, communications and logistics, as per LTG Nutting's concept of technical support for the ESAF. This initial group was to be followed by an additional 250 advisers, who were to be posted in every brigade, battalion, and company in the Salvadoran army. The proposal included the assessment that implementation of this advisory plan would bring the Salvadoran Civil War to termination, on the U.S. Government's terms, in three to five years.[19]

SOUTHCOM accepted the fifty-five-man package, but rejected the 250-man follow-on element because it would have "no political viability or support." The planners had no idea that the fifty-five-man limit would become the ceiling.[20] But without an overall strategy in place, the training was piecemeal and El Salvador Armed Forces (ESAF) commanders selectively obeyed or disregarded their commanders on training.

The election of Ronald Reagan as president of the United States in November 1980 would have profound consequences on the situation not only in El Salvador but across the region. In El Salvador, the FMLN decided to go for broke with the 1981 Final Offensive in hopes of knocking out the government before Reagan took office. Despite having U.S. military aid cut off in early December 1980 due to the killing of four American churchwomen, the Carter administration restored it four days after the launch of the Final Offensive and the government in San Salvador was able to fend off the assault, if barely. Six days later President Reagan took office. Rolling back the advances of Soviet client states would be a key tenet of his early foreign policy.[21]

But even though President Reagan called for a more muscular foreign policy the specter of Vietnam constantly asserted itself into matters relating to U.S. policy toward Central America in general and El Salvador in particular. The fear of U.S. casualties ensured that

> the U.S. MILGROUP [military group]'s Rules of Engagement restricted the trainers to observing and conducting training and activities within five kilometers of their assigned *cuarteles* [garrisons], although they could travel throughout their zone of responsibility if they were escorting a senior ESAF officer. The U.S. trainers were forbidden to accompany ESAF troops into combat, with or without a senior military escort.[22]

Enforcing human rights standards was difficult because trainers could not accompany ESAF on combat operations out of fear of U.S. casualties and loss of congressional support, but such accompaniment would allow for stricter enforcement.[23]

In late 1981, LTG Nutting sent his land component commander—and later SOUTHCOM CINC—then brigadier general Fred E. Woerner to El Salvador to develop a plan to expand and professionalize the ESAF.[24] This report initiated the training of ESAF immediate reaction battalions capable of fighting day and night to put the FMLN on their heels.[25] But in order to comply with the fifty-five-man trainer rule the eventual five immediate reaction battalions (all fielded by 1984) had to be trained, in whole or in part, in El Salvador; Panama; Ft. Bragg, NC; Venezuela; and Honduras.

After all of these battalions were fielded, the operational planning and assistance training teams (OPATTs) really took off in terms of embedding with ESAF brigades. Figure 3.1 shows the disposition of OPATTs at an indeterminate time during the campaign. Retired SF officer and OPATT member Cecil Bailey has stated that from 1984 to 1992 OPATTs rotated through the brigades and extended the progress of the previous teams. "Their achievements and successes must be judged by the cumulative effect of their influence on increasing ESAF counterinsurgency capabilities over time and especially on improving the ESAF's human-rights record."[26]

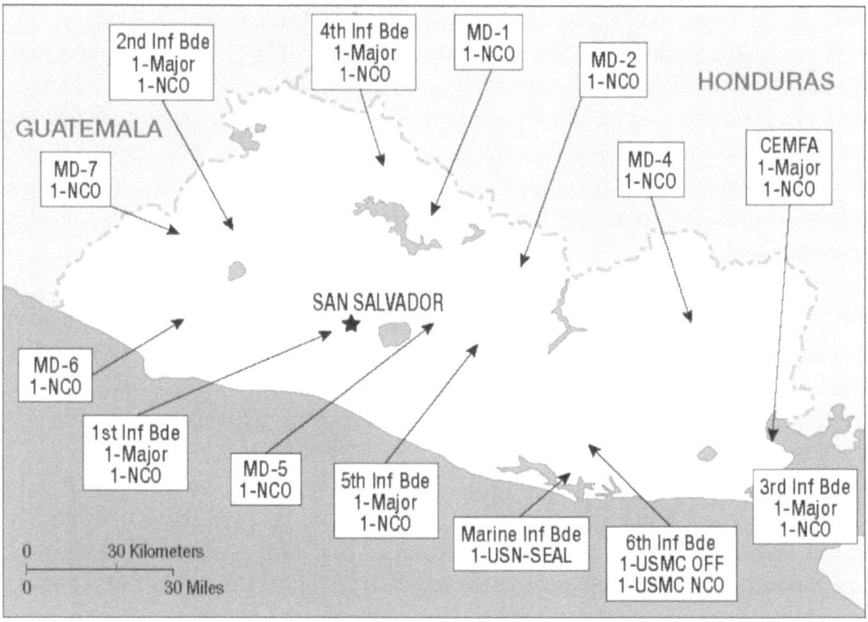

Figure 3.1. A Snapshot of American Advisors with the ESAF. *Source*: Chris L. Lukasevich, U.S. Training and Advisory Assistance to the Armed Forces of El Salvador from 1981–1991 and the Resulting Decline in Human Rights Abuses: The Role of U.S. Army Special Forces (MA Thesis, University of Arizona, 2002), 36.

While some of these advisors were from the U.S. Marine Corps or were Navy SEALs, Sepp noted:

> Almost all the trainers were Special Forces officers and sergeants, who shared a doctrinal language and understanding gained from focused counter-insurgency schooling which emphasized self-reliance, creativity, and working with no resources other than what might be obtained locally. This laissez-faire method of problem-solving was exemplified by the MILGROUP commander's instructions to Captain Craig Leeker on his arrival in El Salvador: "Go to San Francisco de Gotera, and do what you think needs to be done. If you screw up, you'll be thrown out of the country, See you in thirty days." The MILGROUP relied on core Special Forces training and common sense in their trainers to produce the desired results.[27]

Another veteran of El Salvador, Colonel J. S. Roach contends that the OPATTs worked efficiently when they developed rapport and helped brigade commanders accomplish their agendas.[28]

By the spring of 1983, Congress began debating placing the fifty-five-man limit in place by law. In August of that year President Reagan responded to the congressional maneuvering and told them that he would not raise the trainers number above fifty-five. A gentlemen's agreement had been struck. But what constituted this fifty-five number? After discussions with congressional staffers, it was determined that MILGROUP staff, medical trainers, and U.S. military personnel on temporary duty performing service functions (such as communications and helicopter support) would not count toward the fifty-five number. Members of the MTTs were not counted against the fifty-five-trainer cap unless they were in El Salvador for more than fourteen consecutive days.[29]

As stated earlier, the advisors fell under the purview and control of the MILGROUP. According to Colonel James J. Steele, its commander from 1984 to 1986, the MILGROUP "provides basically three things. We provide materiel, we provide training, and we provide advice. We do that consistent with the policy of the U.S. Government as implemented by the Ambassador."[30]

Although prohibited by the rules of engagement from the MILGROUP to accompany ESAF units in the field, some advisors ignored these restrictions. According to Cecil Bailey, "Isolated from anyone who could enforce the rule, they would have a problem only if something went wrong. The joke among advisers who skirted the rule was, Just remember to tell someone, 'If I get killed, drag my body back to the cuartel, call the Embassy and report a terrible training accident.'"[31] Those caught violating these rules, however, were removed from the country.[32]

The advisors also worked closely with other U.S. interagency partners like the AID. Early on historical distrust and friction existed in El Salvador between the interagency and the military, mainly due to Vietnam.[33] Personalities, as well as bureaucratic styles, also hindered relations. As retired Special Forces lieutenant colonel Bob Watson, a former 1st Brigade OPATT leader, has stated, the SF guys were more concerned about going where the action was while the AID personnel were more concerned with taking care of their programs.[34] Over time, the interagency actors achieved unity of effort in El Salvador because of "a function of senior leadership, trust, shared experiences, and the nature of the conflict itself."[35]

Respect for human rights was emphasized throughout training. As Colonel Roach recounted years after the fact:

> Human rights wasn't a separate one-hour block at the beginning of the day. You had to find a way to couch it in the training so that it wasn't just a moralistic approach. It was not an issue that the Salvadorans were confused about. The Salvadoran military understood they weren't supposed to violate human rights, but they believed they were driven to extreme measures by extreme circumstances. When you could convince them in an operational context why human rights made sense, you started to get their attention. When the Salvadorans started to see, "that

if we don't shoot the civilians, we'll get their support, and therefore be able to rout out the guerrillas," then they saw human rights from a pragmatic perspective.[36]

Using data from the UNSC's March 1993 report *From Madness to Hope: The 12 Year War in El Salvador: Report of the Commission on the Truth for El Salvador*, Chris L. Lukasevich demonstrated that "from 1980–1990, direct source *graves hechos de violencia* [serious acts of violence] attributed to the Salvadoran Army declined from 1196/1000 soldiers to 45/1000 soldiers, a 2600% decrease!"[37]

Retired Special Forces colonel and former El Salvador MILGROUP commander John Waghelstein explained that while the fifty-five-man trainer limit caused some problems with neighbors and allies (particularly with Honduras and Venezuela), "it also had benefits, it allowed the Ambassador and military group commander to shunt aside projects they were not enthralled with under a convenient excuse; it forced the Salvadorans to prioritize their wants; and the limit limited US response and helped keep the war a Salvadoran effort."[38] Writing on a different occasion, he offered the following anecdote about establishing leverage with the HN client:

> One such opportunity arose during a conversation with the ESAF Chief of Staff, Colonel Rafael Flores-Lima. While I was trying to make the linkage between reform and Congressional appropriations clear, he reminded me that El Salvador's fight was the United States' fight. I responded that, unlike Vietnam, where we'd committed 450,000 troops, it would not take long to put the 55 trainers on an airplane. I think he understood the tenuousness of our support and the need for the ESAF to do some serious house cleaning.[39]

Furthermore, he added, U.S. aid was leverage and our entry fee but that Congress and a lack of improvement on their part risked the discontinuation of support. The United States could not make support unconditional and had to be prepared to walk away if necessary.[40] Of course, he also argued that U.S. advisors must be culturally and historically in tune with the particulars of El Salvador.[41]

While the specter of Vietnam constantly weighed on U.S. involvement politically and in the media, the interesting thing was the lack of wide-scale panic on the rare occasions when there were U.S. troop fatalities. For instance, when a trainer, Staff Sergeant Gregory Fronius, was killed when the base of El Paraiso was overrun in an attack on March 31, 1987, SOUTHCOM anticipated a storm of protest, but the story went away in a week.[42] A not insignificant number of soldiers earned Combat Infantryman Badges and Combat Medic Badges during the conflict.[43]

The above having been said, not all observers were as sanguine about the performance of the American advisory effort. Robert D. Ramsey III bluntly contends that "the goal of professionalizing the ESAF—of changing its internal values, customs, and traditions to those resembling a modern professional military—did not happen."[44] Similarly, the so-called four colonels report claimed that the advisory effort was performed suboptimally and the right individuals were not sent to do the job.[45] Too rapid personnel turnover, according to some, such as Ambassador Thomas Pickering, also caused effectiveness problems.[46]

At the theater strategy level, LTG Nutting left SOUTHCOM in 1983. Instead of retiring, he earned a fourth star and took over the U.S. Readiness Command in Georgia—the Continental U.S. force provider for the CINCs. During his tenure at SOUTHCOM, he pushed the lines of his strategy based upon bringing stability and democracy to the region.[47] His replacement, GEN Paul Gorman, came into command with his fourth star and with a mandate to escalate the campaign against the Sandinistas. In order to do this, he doubled the size of his staff, greatly expanded communications and intelligence networks, and devised a system using U.S.-based forces to conduct large-scale training operations on the borders and coasts of Nicaragua.

GEN Gorman recognized that the unified and specified commands were not just for deterring war, but also useful for the daily formulation and conduct of foreign policy.[48] Using his extensive networks of contacts, he was able to sidetrack normal channels to go to senior members of DOS, DOD, CIA, and the services to get support and resources as needed.[49] With particular respect to El Salvador, he worked in tandem with Ambassador Thomas Pickering to confront the Salvadoran Minister of Defense and his principals about corruption and diversion of aid. Of his daily briefings at SOUTHCOM, a full 80 percent focused on El Salvador. According to Sepp, "a briefing team dedicated to this effort reported the ongoing activities of every ESAF brigade, and recounted every tactical engagement in the entire country since the previous day, down to squad level."[50]

GEN John Galvin and then GEN Frederick Woerner followed GEN Gorman as the SOUTHCOM CINCs. While the Civil War in El Salvador was stalemated, other events in the region heated up—particularly vis-à-vis Nicaragua and Panama. On GEN Woerner's watch, for instance, "The day after Alfredo Cristiani was elected president of El Salvador on March 19, 1989, under the close scrutiny and with the approval of international election monitors, [he] demoted El Salvador from SOUTHCOM's first priority country to third."[51] But differences between the regionalist GEN Woerner's policy preferences with the newly elected President George H. W. Bush's policy guidance toward Panama's strongman nuisance Manuel Noriega led him to be relieved of command and replaced by GEN Maxwell Thurman on the eve of Operation Just Cause—the U.S. invasion of Panama.

Thurman and his staff would have their hands full between the preparations for Just Cause, the FMLN's second Final Offensive of November 1989, and the murder of six Jesuit priests that same month at Central American

University. The United States made two things clear to the El Salvadorans: (1) we were not abandoning them and (2) peace negotiations, judicial restructuring, and military reform had to be pressed.[52] Between 1980 and 1989 the ESAF had received $958 million in military aid from the United States. With this amount of aid, as one can imagine, there were many allegations of corruption and misappropriation by the ESAF.[53]

When GEN George Joulwan took charge of SOUTHCOM, the FMLN continued to challenge the ESAF with shoulder-fired anti-aircraft missiles. The GEN response with a "logistics surge" to replace ESAF aircraft, along with fuel and ammunition, in three months rather than one to two years helped convince the rebels to accept reasonable conditions to end the conflict. Even the election of Violeta Chamoro as president of Nicaragua did not stem the flow of FMLN resupply. Chamoro's coalition government had out of political necessity kept on Daniel Ortega's brother, Humberto Ortega, as the defense minister. The Nicaraguan military therefore used the summer of 1990 to replace all of the FMLN equipment losses from the failed November 1989 "Final Offensive."[54]

The stalemate and changing geopolitical situation, along with a lot of finesse, allowed the United States to persuade both sides to reach an accommodation. GEN Joulwan for his part has stated that his MILGROUP-El Salvador commander, Colonel Mark Hamilton, had "won the war in El Salvador. There were a lot of people involved; Mark was crucial. He carried out what I wanted. He had to do things that were high risk. He went out with the Ambassador. He called me and I said 'Do it.' He went out with the rebels. . . . They had machine guns pointed right at him during the whole negotiation." In addition, during the peace negotiations of 1991, a senior FMLN guerrilla chief told Lieutenant Colonel Frank Pedrozo, the MILGROUP operations officer, as follows:

> The effectiveness of what we did in El Salvador I think is underlined by two things, the first thing was that the guerrillas said, at the end of the war, in a conversation with a national-level guerrilla leader that I was privy to, that the most damaging thing that occurred in the war was putting American trainers in the brigades. The reason was because as the Army became more professional, they had fewer human rights abuses, and when they had fewer human-rights abuses, the guerrillas lost a lot of their propaganda value and a lot of their recruiting capability, simply because the Americans were changing things.[55]

The crisis finally ended with a peace treaty in 1992.

Philippines, 2001–Present

The United States and the Republic of the Philippines share a complicated history. Captured from Spain at the end of the Spanish-American War of 1898, a long, brutal COIN campaign was needed to subdue it and make it America's first, and only, colony. In 1935, it was granted self-governing commonwealth status. In 1942, the islands fell to Japanese invasion. From 1944 to 1945 Filipinos fought alongside the United States to regain control. In 1946, it gained formal independence. Since independence, it has faced internal threats from groups such as the Huk Rebellion (1942–1954), the communist New People's Army, and numerous Muslim rebel separatists in Mindanao and its southern islands. The end of the Cold War and Filipino domestic political issues concerning national sovereignty led the United States to depart from its large Asian bases of Clark Air Base in 1991 and Subic Bay Naval Base in 1992.

A country slightly larger than the state of Arizona, the Philippines has the world's twelfth largest population (2011 estimate 103,775,002 with a population growth rate of 1.873 percent). According to its 2000 census, the ethnic composition consisted of Tagalog 28.1 percent, Cebuano 13.1 percent, Ilocano 9 percent, Bisaya/Binisaya 7.6 percent, Hiligaynon Ilonggo 7.5 percent, Bikol 6 percent, Waray 3.4 percent, other 25.3 percent. The largest religious denominations are Catholics (82.9 percent) followed by Muslims (5 percent), other Christians (4.5 percent), Evangelicals (2.8 percent), Iglesia ni Kristo (2.3 percent), other (1.8 percent), unspecified (0.6 percent), and none (0.1 percent). GDP per capita, by 2011 estimates, stood at $4,100.[56] The minority status and the Muslim population, located mainly in the south, and other historical grievances caused tensions.

Even prior to 9/11, there were many linkages between the Philippines and international terrorism. Most well known, of course, is the Bojinka plot planned and conceived of by Khaled Sheikh Mohammed (the mastermind of 9/11) and Ramzi Yousef (the first World Trade Center bomber in 1993). Bojinka called for the assassination of the late Pope John Paul II at the World Youth Day to be held in the Philippines in 1995 and for the near simultaneous detonation of twelve international passenger aircraft over the Pacific Ocean. Fortunately, this plan was foiled.

Owing to the lack of economic opportunities, a number of southern Filipino Muslims sought work overseas. For some, this meant work in the Gulf States where their more syncretistic form of Islam came into contact with more orthodox forms. Some Filipino Muslims even fought against the Soviets in Afghanistan. In particular, the Abu Sayyaf Group (ASG) stood out as a group to be concerned about. ASG was formed between 1990 and 1992 by a teacher, Abdurajak Janjalani, who was a member of the Moro Islamic Liberation Front (MILF) and had fought against the Soviets in Afghanistan. The name Abu Sayyaf was Janjalani's *nom de guerre* taken from a famous Afghan mujahedeen named Abdul Rasul Sayyaf. After Janjalani was killed in a gunfight with the members of the Armed Forces of the Philippines (AFP) in 1998, the group turned to hostage-taking and extortion-themed attacks.[57] As will be seen, just as the Salafist Group for Prayer and Combat (GSCP)/AQIM took advantage of the under-governed spaces of the Sahara and Sahel, so too does ASG take advantage of the porous borders of Sulawesi Sea area of the southern Philippines that bordered on Indonesia and Malaysia.

The August 2000 kidnapping of a U.S. citizen, Jeffrey Schilling, brought matters to a head. In September 2000, then admiral Dennis C. Blair, the commander of PACOM,[58] along with officers from the Special Operations Command, Pacific (SOCPAC), traveled to Manila to brief leaders of the Government of the Republic of the Philippines (GRP) and the AFP on the MTT concept to train and equip a company-sized unit to respond to an increase in the level of terrorist threats. Then president Joseph Estrada rejected the American offer of assistance, but when he was forced out of office in January 2001, his replacement, Gloria Macapagal Arroyo, was more amenable to PACOM's offer of assistance.[59]

From February to July 2001, an SF unit trained and equipped a Philippine light reaction company (LRC) whose members had been selected from Philippine SF and scout ranger units to establish a national counterterrorist unit.[60] During training three more U.S. citizens were taken hostage by ASG. Fighting off pressure to deploy the unit before training was complete, the LRC deployed to the south in days after it graduated from training. Unfortunately, however, while the team was deemed proficient, the AFP did not have the proper command and control structure to properly employ the LRC for optimal operational integration. American SF advisors requested permission to accompany the unit, but SOCPAC would only approve a follow-on assessment.[61]

The SOCPAC Commander,[62] Brigadier General Donald Wurster, and the 1st Special Forces Group (Airborne) commander, then colonel David Fridovich, had a meeting scheduled for September 11, 2001, to discuss the issue of command and control over the LRC and on scheduling a terrorist coordination assistance visit (TCAV) as part of the initial-planning survey for Exercise Balikatan 02–1. As one can imagine, "the meeting's emphasis shifted abruptly, however, with that day's terrorist attacks in the U.S. The SOCPAC staff quickly focused on identifying al-Qaeda connections in the region in order to help PACOM develop its Southeast Asian GWOT campaign plan."

The TCAV was planned by Colonel Fridovich and deployed in October 2001.[63] This team included "operations, plans and intelligence officers and one Special Forces ODA and a [Joint U.S. Military Assistance Group (JUSMAG)] representative to conduct a top to bottom, strategic to tactical level assessment of the security challenges concerning the terrorist threats in the Southern Philippines."[64] Coordinating with the U.S. embassy in Manila, the TCAV were given unprecedented access to senior AFP leaders and planning documents, and also visited myriad command levels in Mindanao and Basilan.[65]

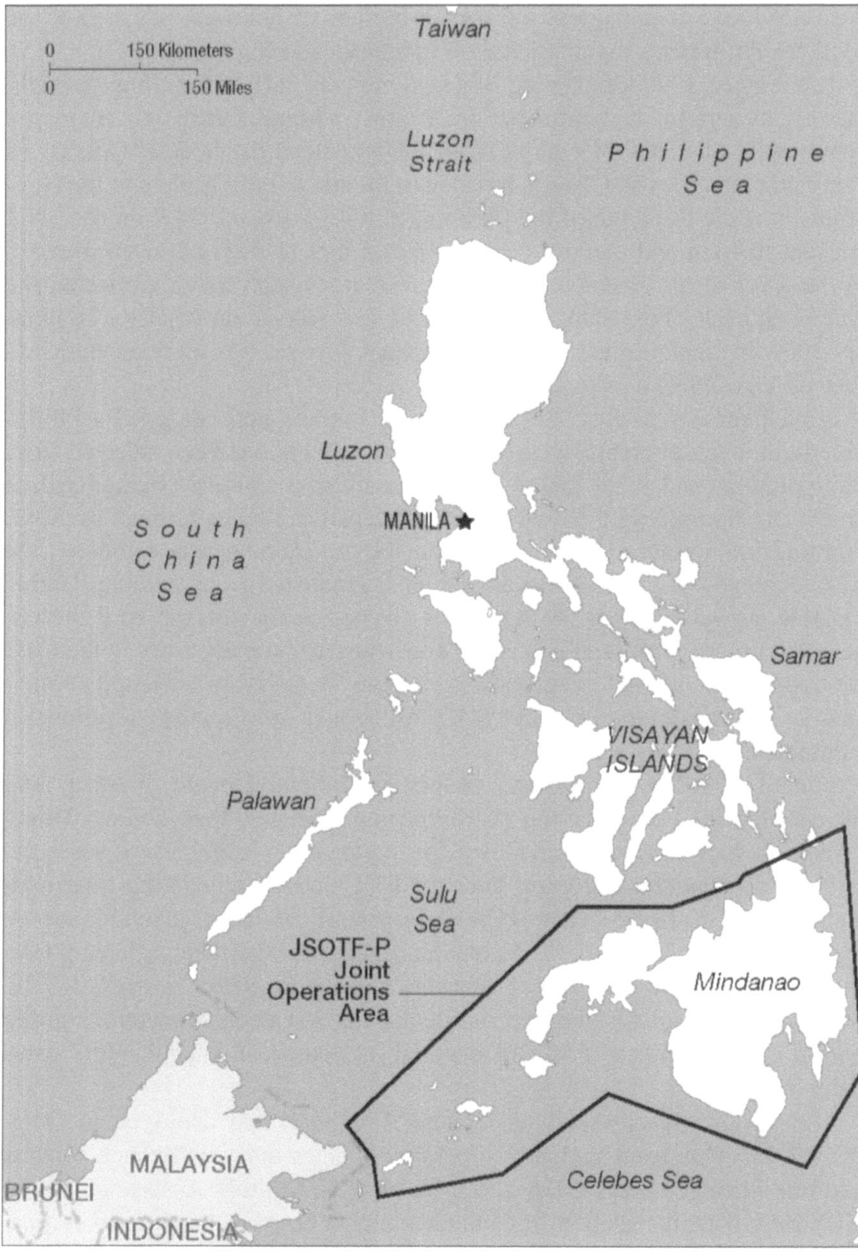

Figure 3.2. JSOTF-P Joint Operations Area.

In November 2001, fighters in support of the Moro National Liberation Front (MNLF) declared war on the GRP after President Arroyo removed the MNLF's leader from leadership over the Autonomous Muslim Region of Mindanao. In response to this violence, PACOM activated its standing JTF-510 to

> plan and prepare to implement the first phase of Operation Freedom Eagle as part of Exercise Balikatan 02–1. In deference to the Philippines' sensitivity to sovereignty issues, the 1st Battalion, 1st SF Group, modified its initial plan. Rather than use Exercise Balikatan as a springboard for conducting a combined unconventional-warfare, or UW, campaign against terrorism in the Philippines, U.S. forces would advise and assist the AFP in IDAD.

The SOCPAC and PACOM staffs labeled their effort the GWOT Southeast Asian Campaign, a name that connoted their regional and long-term involvement.[66]

According to Colonel (ret.) David Maxwell who was commander of the 1/1 SFG(A) at the time, and later the commander of the Joint Special Operations Task Force-Philippines (JSOTF-P), the assessment and plan development noted that (1) the AFP and Philippine government would benefit from the advice and assistance of the U.S. military and other USG agencies, (2) the U.S. forces had to work through and with AFP forces with the Filipinos and no unilateral U.S. combat operations, (3) U.S. forces respect the sovereignty of the Philippines and have an understanding of the environment, and (4) "the U.S. effort had to be joint and interagency and capable of being sustained for a long duration (years), and this in turn required continuous assessment."[67]

Furthermore, he continued,

> Direct FID was the appropriate mission for the U.S. military in the Philippines because the political sensitivities within the Philippines demanded that U.S. forces not be engaged in direct combat operations. While enemy forces have engaged U.S. forces and a number of casualties have resulted, U.S. forces have never conducted deliberate direct combat operations either unilaterally or with Philippine forces. An example of the sensitivity of the presence of U.S. forces in the conflict areas on Mindanao occurred in naming the first deployment of U.S. forces to Basilan in 2002. During operational planning discussions between SOCPAC and the AFP it was decided to call the first deployment "Balikatan" (in Tagalog meaning "shoulder to shoulder") in order to alleviate the concerns of the population. Balikatan is an annual combined exercise normally conducted on the island of Luzon that includes training in combat operations as well as various training scenarios to improve interoperability between the AFP and the U.S. military and these include disaster relief, humanitarian assistance, and civil-military operations. At the beginning of Balikatan the commander of SOCPAC, then Brig Gen Donald Wurster (later Lt Gen) stated that U.S. forces were on Basilan to assist the AFP to defend Filipinos from terrorists and to assist the government of the Philippines with local development projects. One of the

> most important results of this first deployment was described by then-COL Fridovich is that the U.S. forces kept their promise. They did not engage in deliberate direct combat operations, they improved AFP forces, and they supported the AFP in conducting civic action and humanitarian assistance for the local population. Although there was and continues to be vocal political opposition in the Philippines among some segments of the population, the establishment of trust between U.S. forces and the people continues to serve the mission well.[68]

When JTF-510 stood down and transferred control over operations to the JSOTF-P, this also marked the transition from Operations in the Philippines moving from Title 10 to Title 22 security-assistance funding and "control of the missions also passed to the . . . JUSMAG . . . in Manila. The JUSMAG supervised, but since it was located at the U.S. Embassy in Manila, the JSOTF-P had tactical control over the Army SF teams in the joint operating area."[69]

According to Maxwell, "Arguably, JSOTF-P was the right sized force for the right mission based on a thorough assessment of the political and security conditions."[70] At the heart of its success laid its crystal clear mission statement:

> JSOTF-P, ***in coordination with the US Country Team***, builds capacity and strengthens the Republic of the Philippines security forces to defeat selected terrorist organizations in order to protect RP and US citizens and interest from terrorist attack ***while preserving RP sovereignty***.[71]

"When dealing with sovereign countries, the tip of the spear is the Country Team," he has stated elsewhere.[72]

The fact that JSOTF-P[73] had only one Country Team to deal with certainly helps as compared with the next case study on OEF-TS. The development and maintenance of cumulative trusting relationships not only with the embassy but also with the GRP and AFP paid dividends. The JSOTF-P built upon, drew upon, and established relationships over the years between SOCPAC and the U.S. Country Team.[74]

The issue of sovereignty was of key significance in the Philippines because of the special relationship between the two countries. However, respect for sovereignty must be a consideration in all operations by U.S. military forces operating through and with friends, partners, and allies to achieve mutual objectives. Because of the colonial history, U.S. bases on Philippine soil, and political opponents, it was paramount that criticism of the Philippine government and military be deflected or at least not be a result of U.S. actions. Every member of the JSOTF had to be fully aware that every action they took could be under intense scrutiny. In fact, many mistakes were made as military personnel tried to balance security and protection of the force with protection of the mission. However, in these cases coordinated efforts among the Philippine military and government, the U.S. Country Team, and the JSOTF were usually effective in diffusing the situation. Most importantly, the foundation of trust built on the first Balikatan and successive commanders ensuring that the JSOTF remained true to the established principle of trust was critical in being able to diffuse tense situations. Last, the concept of respect for sovereignty was a key component in the operationalization of the lines of effort.[75]

Other benefits of the Direct FID approach of OEF-P were that it allowed decentralized and tailored approach to working with the AFP,[76] showed, as in El Salvador, that a lack of resources could be a positive,[77] and helped to further AFP legitimacy. On the last point, the words of AFP major general Juancho Sabban, commander of Task Force Comet, are instructive:

> For three decades we were using a strategy of force. It turned out to be a vicious cycle. We would have body count syndrome. Commanders would become popular because they were warrior-like. But I saw the more we destroyed, the more the number of enemy increased. There were so many instances of collateral damage and innocent lives being sacrificed. Just by passing through fields with so many battalions we were already stomping on crops and that makes people resent the military. In the course of a firefight school buildings would get burned, houses would be razed to the ground, civilians caught in the crossfire. Everything was blamed on the military.[78]

To be sure, not everything went perfectly. Early on, language skills were a problem[79] and some complained that advisors were unduly burdened with restrictions about what tactical levels they could embed with the AFP. Some claimed that not being embedded at lower levels suboptimized both their effectiveness and legitimacy with their AFP counterparts and ultimately the assistance they could offer to those counterparts.[80]

In October 2012, the GRP signed a peace agreement with the MILF.[81] But conflict has remained. That year also saw the creation of the Maute Group, later known as the Islamic State of Lanao, that has used criminal activities, such as protection rackets, to fund its operations. Meanwhile, the election of President Rodrigo Duterte had, on one hand, somewhat eased tensions with the Moro groups, but also antagonized relations with the United States and escalated attacks on the Maute Group, including the large-scale urban battle in Marawi in 2017.

OEF-TS, 2001–Present

On the continent of Africa, the Sahel region marks the transition from the vast Sahara Desert in the north to the lush savannah in the south. This transition line stretches from the North Atlantic in the west to the Red Sea in the east.

The analysis that follows focuses on the western Sahel nations of Mauritania, Mali, Niger, and Chad. Each gained independence from France in 1960.

As the strategic analyst and author Robert Kaplan has pointed out:

> The countries of the Sahel—which runs through Senegal, Mauritania, Mali, Burkina Faso, Niger, Nigeria, Chad, and Sudan—are among the world's poorest and most unstable, with some of the highest fertility and lowest quality of life anywhere. Governments have little control beyond their capital cities, and throughout the region are many of the ingredients that breed terrorists and their sympathizers: a population disillusioned with its political leadership; a dangerously high number of unemployed young men; Islamic orthodoxy on the rise. Sahelian Africa provides the two conditions essential for penetration by al-Qaeda and its offshoots: weak institutions and the cultural access afforded by an Islamic setting.[82]

Long, porous borders also contribute to instability and make ensuring positive control over national sovereignty difficult. In the words of a State Department CT official:

> The immense size of these countries, their physical geography combined with weak central authority, and the traditional independence of nomadic life styles, make border control and law enforcement exceedingly difficult. No longer isolated from the rest of the world, the traditional caravan routes in this region now serve as conduits for illegal migration and drugs and arms trafficking, as well as a hideout and staging areas for international and regional terrorists and criminals.[83]

Ethnic tensions have also exacerbated problems in Mauritania (between Moors and non-Moors), in Mali and Niger (with Tuaregs[84] in the north), and Chad.

As will be discussed later, the Algerian civil war in the 1990s would have spillover effect for the region. In January 1992, the military canceled the second round of legislative elections after the Islamist party, the Islamic Salvation Front (FIS), won the first round. The military feared that the FIS was staging a democratic coup with only one election. The Armed Islamic Group (GIA) and others rose up to fight. The nasty civil war ended in 1999 with many members of the GIA accepting amnesty. The fighting is estimated to have killed between 150,000 and 200,000 Algerians. One of the other groups to emerge during this fighting would be the Salafist Group for Prayer and Combat (GSPC). Like the GIA, some members of the GSPC cut their teeth as mujahedeen fighting the Soviet Union in Afghanistan during the 1980s.[85] Unlike the GIA, however, the GSPC did not stop fighting against the Algerian state in 1999.

The attacks on the U.S. homeland on September 11, 2001, with good reason made U.S. decision makers nervous about the possibility of threats metastasizing in areas beyond Afghanistan. The lax border controls, involvement of individuals such as Abu Hafs al Mauritani in key leadership positions in al-Qaeda, and concerns over the spread of the GWOT led some in Washington to think about engagement in the region. As William J. Foltz, an African specialist at Yale University and former member of the National Intelligence Council, stated in an interview, "If you take the whole Sahel, it's not a threat in the sense that it's going to declare war on the United States, but the United States generally pays a price if there is great disorder going on in some part of the world."[86]

Following this rationale, the State Department began the Pan Sahel Initiative (PSI) in November of 2002; it existed until 2004. According to the reputable Globalsecurity.org website, the PSI was a State Department–led effort designed to protect borders, track the movement of people, combat terrorism, and strengthen regional cooperation and stability across Mali, Niger, Chad, and Mauritania. Its goals supported the U.S. national security issues in Africa of combating terror and enhancing regional peace and security.[87] Funding of $6.25 million was provided for the first year and $7.5 million was allotted for the second year.

At the time, the U.S. European Command (EUCOM), as the GCC, held responsibility for the Sahel region. EUCOM in turn directed its subordinate Special Operations Command, Europe (SOCEUR) to "develop and maintain a flexible, enduring Global War on Terrorism Coalition, as well as regional, operationally-focused partnerships to support specific Global War on Terrorism operations."[88]

Key tasks for the main effort included the following: enabling other countries to effectively combat terrorists in their region; improving PN capabilities to support GWOT-related coalition operations; denying safe havens to terrorists in the EUCOM area of responsibility (AOR); and assuring access throughout the AOR for basing and logistics to support operations. The supporting effort also included a number of key tasks: finding al-Qaeda and other terrorists, their networks, and potential safe havens; capturing or killing known terrorists in the EUCOM AOR; and neutralizing known terrorist networks operating in the EUCOM AOR.[89]

The military component of PSI would eventually become OEF-TS. SF and Marine Corps personnel would be sent to Mauritania, Mali, Niger, and Chad in order to train company-sized formations in marksmanship, patrolling, and communications skills, and to partially equip these units to counter terrorist threats in their countries. Table 3.2 shows the manning levels and roughly sketches the capabilities of security forces in the western Sahel in 2002.

In 2003, elements of the GSPC would get daring. That year a leading figure within the group, and former Algerian army paratrooper, Ammari "El Para" Saifi kidnapped thirty-two Europeans (from Germany, Austria, Switzerland, and the Netherlands) in southern Algeria, held them for 177 days, and then ransomed them for €5 million.[90] He then snuck across the border into neighboring Mali. There, according to European CT officials, Saifi bought protection from local tribes, purchased weapons and trucks, and expanded his militia to include fresh recruits from Mali, Chad, and Niger. "At this point, he starts to become an important threat," said Louis Caprioli, former director of

Table 3.2. Western Sahel Security Forces, 2002[1]

	Mauritania	*Mali*	*Niger*	*Chad*
Army	15,000	7,350	5,200	~30,000 (including 5,000 Republican Guards)
Navy	~500 (7 patrol craft)	50 (3 river patrol craft)	N/A	N/A
Air Force	250 (8 combat aircraft)	400 (16 combat aircraft)	100 (no combat aircraft)	350 (2 fixed-wing combat aircraft, 2 armed helicopters)
Paramilitary	~5,000	4,800	5,400	4,500
Foreign Forces	N/A	N/A	N/A	950 French troops, 5 combat aircraft

[1] Data for this table were derived from International Institute for Strategic Studies, *The Military Balance 2002–2003* (London, UK: Oxford University Press, 2002).

international CT for the DST, the French counterintelligence service, which was working with the U.S. military to track Saifi. "And the armies of Niger and Mali couldn't do anything about him."[91]

Shortly after the hostages were released, EUCOM brought the Sahel defense chiefs to its headquarters in Stuttgart, Germany. Reportedly, for the first time in their lives the defense chiefs of Mali and Chad talked. Through this set of face-to-face talks, the PSI "essentially, [laid] the foundation for a network that could stymie the growth of regional terrorism."[92] Ground liaison forces and enabler capabilities such as the P-3 Orion intelligence, surveillance, and reconnaissance (ISR) aircraft were dispatched to help the local forces to track down Saifi and his followers.[93]

Local forces, bolstered by the American capabilities, pursued Saifi's band from Mali, through Niger, and finally to Chad. While U.S. sources have issued denials about the presence of boots on the ground, some sources cite Nigerien officials as claiming that there were American SOF on the ground and were present for at least one gun battle.[94] While events on the ground were not always smooth,[95] Chadian forces with an emergency airlift supply from the United States and with American advisors—and probably French advisors—were able to inflict heavy casualties on the GSPC militants.[96] Rebels of the Movement for Democracy and Justice in Chad captured Saifi. After an awkward attempt to turn him over to the United States, France, and others, the rebels turned him over to Libya, who then extradited him to Algeria to await trial.

Meanwhile in April of 2004, after the battle in Chad, EUCOM convened a meeting of nine African military chiefs again in Stuttgart. While some antagonisms remained between them, each of the nations expressed concern over "the growing threat of Islamic militancy within their borders." U.S. military officials expressed concern that Islamists in the region were able to communicate with one another while governments and their forces were not able to. According to then GEN Charles Wald, GPS equipment allowed militants to create "virtual garrisons" in the desert where they knew where to obtain logistics and store caches of weapons and ammunition.[97] These meetings laid the groundwork for the transition from the PSI to the better-resourced Trans-Saharan Counterterrorism Initiative (TSCTI).

TSCTI officially kicked off in June 2005 with a Flintlock Exercise in Senegal. Joining Mauritania, Mali, Niger, and Chad in the Initiative were Morocco, Algeria, Tunisia, Senegal, Burkina Faso, and Nigeria. The DOD would move from the supported element to a supporting actor to the State Department and other interagency partners, and its contribution would officially become OEF-TS. GEN Charles F. Wald noted the uniqueness of OEF-TS and TSCTI. In his words:

> Operation *Enduring Freedom*—Trans Sahara is an example of operationalized [Theater Security Cooperation (TSC)]. It is the first time a series of TSC events has been grouped under the umbrella of a named operation directed by the Joint Staff. It is the American military component of the Trans-Sahara Counterterrorism Initiative, a long-term U.S. Government program designed to help the countries of Trans-Sahara Africa cooperate to control the undergoverned spaces of their interiors.[98]

While some claimed that EUCOM was overselling the threat posed by AQAM elements in the Sahel,[99] warnings and indicators were pointing to real issues. As Operation Iraqi Freedom became bogged down in a messy insurgency, it began to attract foreign fighters from around the globe. As GEN James L. Jones stated in 2005, "We already have evidence of fighters going from Africa, spending some time in Iraq and Afghanistan, and then migrating back to Africa."[100] On the fifth anniversary of the 9/11 attacks, the GSPC formally merged with al-Qaeda and became AQIM.[101] Furthermore, al-Qaeda in Iraq (AQI) personnel records captured by coalition forces in the northwestern city of Sinjar showed an inflow of foreign fighters from Mauritania and North Africa.[102]

Much like the Arab Afghans who preceded them, the post-2001 foreign fighter veterans of Afghanistan and Iraq were a legitimate concern for the region not only because of their combat experience but also in their ability to recruit others.[103] As retired coordinator for CT at the State Department (with Ambassadorial rank) and Army LTG Dell Dailey and a co-author noted in 2006:

> Terrorists develop active support for a given cause by espousing a message or ideology that resonates with a targeted populace. Tacit support results from some combination of fear and apathy among those not inclined to believe their message. As a popular support base develops, terrorists expand their freedom of action locally or regionally within states that either cannot or will not enforce

> the rule of law. Unwilling or incapable states are vulnerable to becoming havens for terrorist organizations and provide operational freedom of action. The establishment of global network links through a combination of safe haven and operational freedom of action permits terrorists to achieve global reach and project power. Finally, the cycle is completed when successful terrorist operations reinforce ideological justification and influence that portion of the populace that is susceptible to the extremist ideology—bringing new recruits and resources to the cause.[104]

The continued—and continuing[105]—use of hostage-taking and ransoming and involvement in, for instance, the smuggling and protection of illicit cargo, such as drugs,[106] have allowed AQIM to finance itself and avoid some of the counter-threat finance hurdles put in place during the 2000s.[107] The noted defense scholar Frank G. Hoffman has labeled such self-financing as "auto-catalytic operations,"[108] and these fit in nicely with his theory of hybrid threats where multimodal and multinodal operational blurring occurs.[109] Robert Fowler, a former U.N. special envoy to Niger who was held hostage by AQIM for 130 days in 2008–2009, told an audience at Britain's Chatham House think tank in 2011 that his captors told him they wanted to "turn the Sahel into one vast, seething, chaotic Somalia."[110]

A major change occurred in the command and control structure for OEF-TS in 2007 with the arrival of a new GCC with sole responsibility for the continent of Africa (minus Egypt). U.S. Africa Command (AFRICOM) officially took charge of the new AOR with initial operations on October 1, 2007, and officially became an independent command one year later. Its mission statement reads: "U.S. Africa Command protects and defends the national security interests of the United States by strengthening the defense capabilities of African states and regional organizations and, when directed, conducts military operations, in order to deter and defeat transnational threats and to provide a security environment conducive to good governance and development."[111] Like EUCOM, AFRICOM's headquarters is located in Stuttgart, Germany.

Military responsibility for TSCTI shifted from SOCEUR to Special Operations Command, Africa (SOCAFRICA) in 2008. The move also resulted in a name change to the Trans-Sahara Counterterrorism Partnership (TSCTP). The official AFRICOM fact sheet for TSCTP states that it

> is a multi-faceted, multi-year U.S. Government (USG) program aimed at defeating terrorist organizations by:
>
> - strengthening regional counterterrorism capabilities,
> - enhancing and institutionalizing cooperation among the region's security forces,
> - promoting democratic governance,
> - discrediting terrorist ideology, and
> - reinforcing bilateral military ties with the United States.

> The overall goals are to enhance the indigenous capacities of governments in the Pan Sahel (Mauritania, Mali, Chad, and Niger, as well as Nigeria and Senegal) to confront the challenge posed by terrorist organizations in the region. Additionally, TSCTP will facilitate cooperation between the Pan Sahel countries and Maghreb partners (Morocco, Algeria, and Tunisia) in combating terrorism.[112]

The TSCTP budget for FY2007–FY2013 was estimated at roughly $100 million per year.[113]

OEF-TS provides a much different case than both El Salvador and the Philippines. The regional scope, whether one examines PSI, TSCTI, or TSCTP, ensures that the tactical-level view will focus on the involvement of American military personnel acting as part of the various Country Teams. Ambassador Johnnie Carson, the assistant secretary of state for African affairs, testified before the Senate Foreign Relations Committee that representatives from Washington regularly met with embassies in the TSCTP countries and that these embassies were the first line of coordination and that the COMs had to concur with all proposed activities because "they are best placed to understand the immediate and long-term implications of various activities and are ultimately the primary interlocutors with the host countries."[114]

Far from simplifying matters, however, this can lead to power squabbles and personality conflicts.

In a paper written at the U.S. Army War College, Foreign Service Officer F. John Bray provides chapter and verse for difficulties arising in relations between military personnel and other interagency actors in embassies in the Sahel.[115] In his analysis, there was a lot of friction in the interagency which occurred in three domains: institutional, cultural, and legal. Expanded military efforts were problematic because the embassies lacked the institutional capacity to support them. The culture of embassies had their own unique rulebooks that were alien to outsiders. Last, while the GCCs were focused on the region, the ambassadors had to manage bilateral relationships with sovereign nations.[116]

Ironically, while the initial PSI was to build HN capacity to counter violent extremism, the planners and powers that be did not analyze the embassies that U.S. trainers would have to work through. Had such staff work taken place it would have revealed that a whole-of-government approach was impossible in 2002 because the embassies had had their support capability stripped away for cost savings. For instance, the U.S. embassy in Mauritania consisted of seven personnel on 9/11: "an ambassador, a deputy chief of mission (who also served as the political officer), an office management specialist, a duel-hatted consular/commercial officer, a communication officer, an administrative officer, and a newly arrived regional security officer." The embassies in Mauritania, Mali, Niger, and Chad were classified as part of the Special Embassy Program and received special relief from mandatory reporting, were locked up at night, could not process top secret information, and all taskings had to be approved by the State Department's undersecretary for management.[117]

Interagency cultural frictions, personality conflicts, issues relating to proper behavior in dealing with the local populace, and ruffled feathers over operating within an embassy culture all caused issues on the ground.[118] More problematic, however, was over different interpretations of authority between the COM and the GCC. According to Bray:

> While the authorities of the GCC and COM are subject to interpretation, they appear to be relatively clear under two situations: (1) During full combat operations authorized by National Command Authority (GGC leads); and (2) Stable situations when there are no combat operations (COM leads). Although DoD had an execute order signed by the Secretary of Defense and coordinated through DOS and the NSC, TSCTP fell somewhere between these two well-defined situations, and the result was interagency friction between DoD personnel and the embassies in the region. This friction largely overshadowed synchronizing activities that occur as national strategy is implemented at the country level.[119]

Another irony of the clash of cultures and personalities is the fact that advisors, particularly those working in cross-cultural settings, are supposed to be able to work through such differences—and this is especially the case for SF soldiers who are carefully screened for maturity, flexibility, and empathy. But just as the conflicts in Afghanistan and Iraq stretched the personnel assignments of foreign service officers and others, so too did those conflicts stretch American forces and particularly SOF. Early on in PSI, the Ft. Carson, Colorado- and Stuttgart, Germany-based 10th Special Forces Group (Airborne) did most of the training iterations and deployments, but that seemingly decreased over time. 10SFG(A) historically had been focused on Europe, but Iraq and Afghanistan made it send teams to work outside of their cultural and linguistic training in order to allow other Iraq and Afghanistan regionally aligned soldiers to rest and recuperate and attend necessary military schooling between combat deployments. Eventually as 10SFG(A) teams were pulled away from more regular PSI/TSCTI/TSCTP deployments, mixes of Marines and members of other SF Groups filled in the gaps.[120]

But that should not to be construed as being hypercritical of American SOF, and army SF in particular. Elements of SOF did many innovative and adaptive things as part of OEF-TS. One such development was the creation in late 2006 of the JSOTF-Trans-Sahara (JSOTF-TS) by a group of SOF officers. JSOTF-TS since then has served "as a unique organization capable of orchestrating counterterrorism objectives under the auspices of the US State Department initiated and led Trans-Sahara Counterterrorism Initiative (TSCTI) with its integrated 3-dimensional strategy of diplomacy, development and defense."[121] Army CA units have also undertaken key development projects to further U.S. objectives in the region.[122]

While some have criticized American involvement in the western Sahel for being too focused on military means,[123] of empowering autocratic rulers,[124] or of planting the seeds of future coups,[125] in the main, the U.S. approach in the region has focused on its concern for international terrorism rather than on domestic security[126] and for getting ahead of potential transnational threats. As Robert Kaplan observed regarding Mali:

> Once again, the point was for the U.S. military to get into a place early, fast, and with a small foot-print to try to do some good, so that it wouldn't need a bigger one later on. Even if the Salafist terrorist threat was exaggerated, much like the Cold War, it served as a useful pretext for military engagement with obscure parts of Africa: something that helped stabilize newly civilian regimes and got relief aid to places otherwise forgotten. It also fostered relationships with local intelligence assets of the kind the American military wished it had had in Afghanistan and elsewhere following 9/11.[127]

Of course, that was in 2007. In 2012, the overthrow of Libyan leader Muammar Qaddafi led to the return of heavily armed Tuareg mercenaries who had fought for his regime to the north of Mali.[128] That March the Malian army launched a coup d'état against the long-standing democratic government over its concerns about the deteriorating security situation in the north of the country.

While the United States suspended military relations with Mali that same month, reporting by the *Washington Post* after a car accident which killed three U.S. SOF personnel when their Toyota Land Cruiser crashed off a bridge into the Niger River in Bamako on April 20, 2012, raised questions about the ongoing CT operations in the region. AFRICOM told the paper that the soldiers had been in the country working with the Malian military prior to the coup and had remained as part of a "small number of personnel" to "provide assistance to the U.S. Embassy" and "maintain situational awareness on the unfolding events." Two of the soldiers, Captain Daniel H. Utley and Sergeant first Class Marciano E. Myrthil, were members of the 95th CA Brigade from Ft. Bragg, NC, and the other, Master Sergeant Trevor J. Bast, was from the Army's Intelligence Security Command, which reportedly conducts signals intelligence and works closely with the Joint Special Operations Command (JSOC).[129]

The *Post* reported that classified intelligence-gathering operations were taking place across the Sahel and, furthermore, that a program named Oasis Enabler was proposed in 2009 that would embed U.S. SOF with Malian ground troops to conduct CT operations. Furthermore, they reported that in an October 2009 meeting between the U.S. Ambassador to Mali, Gillian A. Milovanovic, and the deputy chief of AFRICOM, Vice Admiral Robert

T. Moeller, the ambassador allegedly protested arguing that such a display might anger neighbors and expose the Malian troops and our personnel to unwanted risk. The admiral reportedly replied that he "regretted" that the ambassador had not been kept better informed and said Oasis Enabler was "a work in progress."[130] While it is unclear whether such embeds took place, if they did, then this issue does raise some questions and concerns. Moreover, the security situation in Mali and the spread of Ansar Dine and the presence of AQIM and foreign fighters from the region, as well as reports of some fighters from South and Central Asia, have led the former assistant secretary of defense for Special Operations and LIC, Ambassador Michael Sheehan, to state at a summer 2012 Aspen Institute event that the United States "can't allow al Qaeda to sit in ungoverned space" in northern Mali and that the United States and allied officials had discussed numerous scenarios for dealing with the situation to include "perhaps operating in ungoverned space."[131]

The continued spread of an Islamist Ansar Dine movement in the north of Mali[132] became so destabilizing that the French military intervened to help the government fight it. Army GEN Carter Ham, a former AFRICOM commander, for instance, confirmed that some individuals involved in the attacks on the U.S. consulate and a CIA annex in Benghazi, Libya, had links to AQIM.[133] But the threat expanded with the arrival of the Islamic State and its aligned groups.

On October 4, 2017, four SF soldiers and five Nigerien soldiers were killed in an ambush. While originally going out to conduct a key leader engagement, the SF team was retasked by another junior leader with assisting in an assault to capture Doundoun Cheffou, who was linked to the Islamic State. The team leader "warned before the mission that his troops did not have the equipment or intelligence necessary to carry out a kill-or-capture raid against a local militant."[134] On December 6 of that year, United States and Nigerien forces came under attack by militants again. This time no U.S. or Nigerien forces were killed, but eleven militants were. The DOD quietly revealed that there had been ten such instances of attacks in Niger or other states in West Africa from 2015 to 2017.[135]

By 2018, as the Trump administration had outlined its plans to focus more on GPC, it was reported that AFRICOM was asked how it would deal with CT operations with a reduction in its then assigned 1,300 personnel by 25 percent in eighteen months and by 50 percent over three years.[136] But in the fall of 2018, *The New York Times* reported that there was an expansion of the CIA's use of drones in the Sahel with a new base at Dirkou, Niger.[137] As

Table 3.3. Western Sahel Security Forces, 2019[1]

	Mauritania	*Mali*	*Niger*	*Chad*
Army	15,000	10,000	5,200	~25,000
Navy	~600 (17 coastal and patrol craft)	N/A	N/A	N/A
Air Force	250 (4 combat aircraft)	400 (4 combat aircraft, 4 attack helicopters)	100 (2 combat aircraft, 2 attack helicopters)	350 (14 combat aircraft, 5 attack helicopters)
Paramilitary	~5,000	7,800	5,400	9,500
Foreign Forces	N/A	1,750 French troops	500 French troops, 4 combat aircraft	1,500 French troops, 4 combat aircraft

[1] International Institute for Strategic Studies, *The Military Balance 2019* (London, UK: Taylor & Francis, 2019).

the Rand Corporation's Linda Robinson has stated about the Pan Sahel and Trans-Sahara mission:

> The U.S. undertaking an advisory approach to the Africa threat was, at the outset, a recognition that it wasn't a top-tier threat. It was more of an incipient threat that could become a greater threat to Europe and, in the case of some groups, potentially come to threaten U.S. interests overseas. It was more a concern about the disruption to friends and allies in that region and the idea that these groups could grow. It was a way to try to address these threats while they were still more regional and locally focused.[138]

The projected population growth in Africa and the competition between various militant groups both among themselves and with local governments will likely drive the level of U.S. involvement moving forward.

ASSERTIVE DEFENSIVE INDIRECT INTERVENTIONS

In order to fill in the assertive defensive indirect intervention quadrant, the case of the British intervention in Oman (1965–1975)[139] has been selected. Oman is an interesting case to consider not only because it perfectly illustrates such a small-scale defensive intervention but also because by the way it was carried out it shows how the United Kingdom used such a covert intervention when it was publicly scaling back its strategic commitments. In the future, it may be possible to use the American involvement in Iraq post-2014 to provide an additional example, but there is not enough information on that case at this time to make an in-depth analysis.

Oman lies at the far southeast of the Arabian Peninsula and is slightly smaller than the state of Kansas. It is a remote kingdom that follows the Ibadi school of Islam. Its relations with the Great Britain began with a treaty of friendship in 1798. The British were worried that the French might use Oman as a staging area for an invasion of India. By 1815, British concerns over piracy and Gulf shipping led to naval support to Muscat.[140] But unlike the neighboring United Arab Emirates, formerly the Trucial States, or Aden, it was never a British colony. The country's crucial position on the approach to India made it a valuable friend to the crown. Its position along the trade routes of the Indian Ocean also allowed the Omani Sultan to have territorial possessions as far south as Zanzibar and as far east as the port of Gwadar in Pakistan. The discovery of oil in the Middle East made its position on the Strait of Hormuz a critical geopolitical stakeholder—its noncontiguous Musandam Peninsula sits at the Strait.

Prior to the oil shocks of the 1970s, Oman was an isolated and poor country. As Ian Cobain has pointed out, under the Sultan Said bin Taimur (1910–1972),

> in the mid-1960s, Oman had one hospital. Its infant mortality rate was 75% and life expectancy was around 55 years. There were just three primary schools—which the sultan frequently threatened to close—and no secondary schools. The result of this was that just 5% of the population could read and write. There were no telephones or any other infrastructure, other than a series of ancient water channels. The sultan banned any object that he considered decadent, which meant that Omanis were prevented from possessing radios, from riding bicycles, from playing football, from wearing sunglasses, shoes or trousers, and from using electric pumps in their wells.[141]

Citizens of the sultanate who wanted to receive an education or improve their economic opportunities largely had to do so abroad. And in a Middle East teaming with nationalist and socialist sentiments in the 1960s, other states were happy to try to engage in training and education to help produce cadres that might help to make political changes on the ground. The Soviet Union, China, and countries like Iraq were all too welcome in providing such training to Omani expatriates.[142]

The Dhofar region of Oman, in the southwest part of the country bordering in Southern Arabia (later the People's Democratic Republic of Yemen (PDRY) and later just Yemen) and Saudi Arabia, was geographically, ethnically, and culturally isolated from the capital of Muscat and the rest of the country. Before the 1970s, one road connected the city of Salalah with the rest of Oman. Dhofar contains a lush coastal plain backed by harsh mountains and followed by a moon-like desert. Monsoons from the Indian Ocean affect the region from May to September. The jungles and craggy mountains made for ideal guerrilla territory.

The Dhofar Rebellion started for numerous reasons. The initial impetus for the rebellion was Dhofari dissatisfaction with Sultan Taimur's policies and neglect. In power since 1932, Taimur's approach was to not allow the people to become too accustomed to modernity. While he built a palace in Salalah and married a Dhofari woman, he did little to foster unity with the Dhofari people. He did not allow any kind of agriculture throughout that coastal plain backed by the mountains of the Jebel Dhofar. Oman at this time had a population of about 700,000 people and Dhofar had a population of approximately 50,000. Ten thousand of the Dhofaris lived in the mountains and were referred to as Jebelis. The Dhofaris saw themselves, in some sense, as being occupied by Arabs from the north.

The insurgency began across various strands, some nationalist, others communist, before organizing under the umbrella of Dhofar Liberation Front (DLF) in 1965. The rebellion started out very slowly, going after oil concessions and exploration sites up toward the border with Saudi Arabia and then eventually to active attacks on the Sultan's Armed Forces (SAF). Their goals sought development and aid for the region, as it was underfunded compared to Muscat and the Arab portions of the country.

In 1965, in response to these attacks, the Sultan invited the British in and they did some initial operations from the Trucial States (present-day UAE) going through Dhofar that were small-scale operations of limited duration and effectiveness. (Britain was concerned with upholding the Sultan's government from foreign influence and interference, maintaining access to air bases at Salalah and Masirah and overflight rights, and to safeguarding oil interests.[143]) The SAF at the time were very small, comprising a few thousand soldiers drawn from Arabs in the north and Baluchis from Pakistan. The British Army had seconded[144] officers to the SAF. The commander of the SAF was a British officer, and all the way down to battalion level there were as many as three British Army captains for every company in the SAF, but it was a very small organization. Contracted soldiers, pilots, and support personnel were also in Oman.

But the situation in neighboring Aden and Southern Arabia would have a large impact on the situation on the ground. After the British departed Aden in 1967, communists took power and established the PDRY (also known as South Yemen). The PDRY was supported by the Soviet Union and the People's Republic of China. By 1968, communist elements, supported by Chinese, Soviet, and Yemeni training and equipment, took control of the DLF and transformed it into the Popular Front for the Liberation of the Occupied Arabian Gulf (PFLOAG).[145]

The British involvement in the Dhofar campaign can be conceived of in three phases. Phase I lasted from 1965 to 1970. This period was an abject failure. The wadis and hills of Dhofar made it a particularly difficult area of operations. Phase I attempted an attrition-based strategy without the necessary manpower and equipment. The PFLOAG, on the other hand, had enough numbers and freedom of movement to keep the small numbers of SAF at bay and were much better equipped. While the SAF were equipped with bolt action World War II–era Lee-Enfield rifles and Bren guns, the PFLOAG had modern Kalashnikovs (AK47), RPKs (Kalashnikov light machinegun), and RPGs (rocket-propelled grenades). Only the coastal towns of Salalah, Mirbat, and Taqa were under governmental control by 1970.[146]

Phase II took place between 1970 and 1972. In 1970, the commander of the 22 Special Air Service (SAS) Regiment, looking for work for his regiment, Lieutenant Colonel Johnnie Watts, went to Dhofar and conducted a hasty assessment on the ground coming up with a five-part plan for winning the war there.[147] (Watts had previously served in Oman in 1959 when the SAS had helped battle an uprising by a Saudi-supported Omani imam in the mountains outside of Muscat.[148]) This plan was based on civil reorganization, agriculture and economic development, intelligence gathering, psychological warfare, and military operations and training of local forces. SAS personnel were given Arabic language training and repetitive deployments gave them deep experience on the ground.[149]

Sultan Taimur wanted to keep this operation very low key and did not want an acceleration of British support in Oman. His son, Qaboos bin Said (1940–2020), however, had had enough. Supported by the British, Qaboos launched a coup d'état against his father in the summer of 1970 and took power. The new sultan, whose mother was Dhofari, had trained at Sandhurst and served in the British Army's Cameronian Rifles in Germany. After his service he studied modernization issues in the United Kingdom. With his rise to power, he decided to change the approach on the ground there. The first thing he did was to address the social, political, and economic concerns. He offered general amnesty to all the subjects who opposed his father. He incorporated Dhofar formally into Oman as the southern province and changed the name of the country from Muscat and Oman to the Sultanate of Oman. He decided to provide effective military opposition to the rebels who did not accept the amnesty offer. He started a vigorous nation-wide program of developments, and finally he started diplomatic initiatives with the aim of having Oman be recognized as a genuine Arab state and to isolate the PDRY.

After the coup, Sultan Qaboos more formally invited the 22 SAS to come into Oman. Operating under the cover name of the British Army Training Team (BATT), one squadron of SAS personnel (roughly seventy soldiers) started to filter through to conduct civil action programs, and they began to raise local civilian irregular defense groups called *firqats* (Arabic for company). Some members of the *firqats* came from surrendered enemy personnel. Early on there were concerns that any British casualties would cause a premature end to British involvement.[150] But Sultan Qaboos kept a tight lid on foreign press in Oman, so British involvement was very lightly reported on. As the scholar Walter Ladwig has pointed out:

> Close coordination between the Sultan's Forces and the British government occurred at the country team level. The Commander of the Sultan's Forces met daily with the British embassy's defense attaché, and weekly on an informal basis with the British ambassador. This system was adopted to deflect criticism that the British CSAF meeting daily with the British ambassador was a clear indication that London was really running the show in Oman.[151]

This tracks well with the American cases discussed earlier in this chapter. Furthermore, seconded officers were said to take their duty to the Sultan seriously and even withheld information from London about sensitive operations in case objections might be raised.[152]

The BATT and the *firqats* started to work in the Jebel. These units and the SAF started to implement a population-centric approach. Engagement with the Jebelis led to increased civic action activities. They started to build wells and doing things like cattle inoculation. They undertook a cattle drive to bring cattle to market there. As the population was largely illiterate at the time, transistor radios were subsidized and sold in local markets and provided access to "warts and all" news programming that was contrasted with agitation and propaganda broadcasts from the PDRY.[153] And in combination with the SAF they started to push the supply lines of the PFLOAG back toward Yemen with a series of defensive lines—the Hammer Line, the Hornbeam Line, and Damavand Lines. Contractors and British Army personnel worked alongside the SAF and equipment such as Canberra bombers were used to conduct strike operations against the insurgents.

The turning point in the war occurred in July 1972 with the battle of Mirbat. Mirbat was a town to the northeast of Salalah surrounded by water on two sides and isolated. The PFLOAG gambled that a large raid and temporary seizure of the town from the defenders led by a BATT, *firqat*, and Omani gendarme garrison could have strategic impact and produce a major propaganda victory.[154] On July 18, a force of between 250 and 300 *adoo* attacked the garrison. While seven members of the SAF and two members of the BATT were killed, it is estimated that between 80 and 200 *adoo* were killed. The *adoo* had underestimated the ability of the defenders to receive air support during operations and a squadron of SAS freshly rotating into the country were able to be airlifted in to help turn the tides of this battle.[155]

After the battle, there were increased numbers of surrendered enemy personnel. An SAS officer later remarked that pre-Mirbat the *adoo* had initiated contact 99 percent of the time, while after the BATT, *firqat*, and SAF initiated 50 percent of the time.[156] If accurate, this further seems to reflect a major psychological turning point for the SAF. Sultan Qaboos's diplomatic initiatives

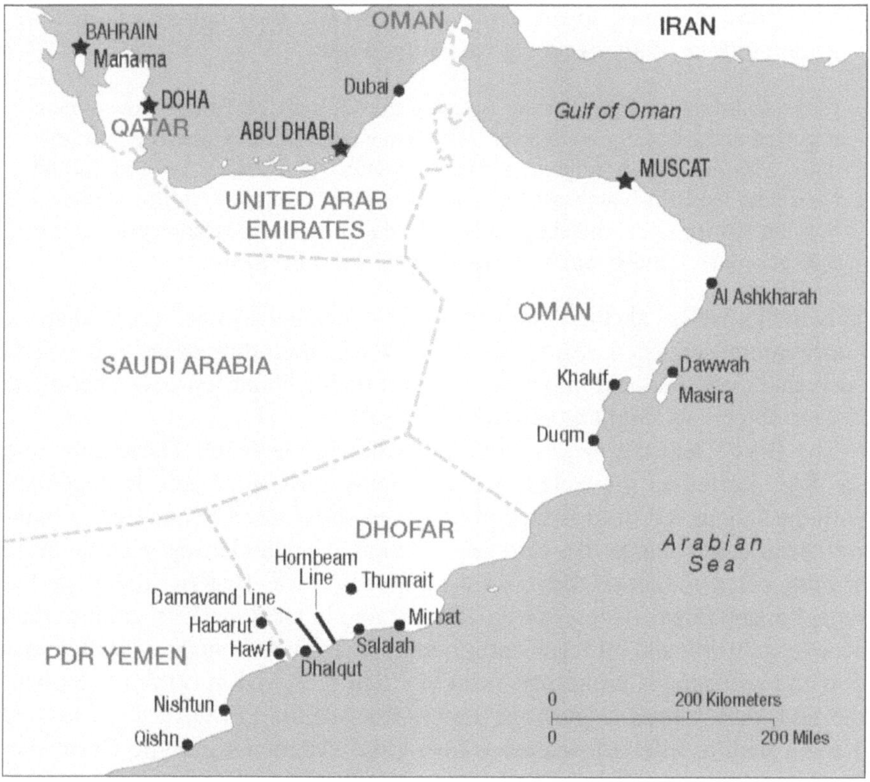

Figure 3.3. Dhofar, 1965–1975. *Source*: Michael P. Noonan, FPRI.

to join the Arab League and the United Nations in 1971 and establish relations with neighbors would also lead to dividends in the next phase of the conflict. (Prior to that, Oman had diplomatic relations only with the United Kingdom, the United States, and Pakistan.[157])

Beginning in November of 1972, the BATT also started to train exiled Mahra tribesmen from the PDRY to conduct cross-border operations.[158] While these operations were billed as taking the fight across the border, it appears that they were principally conducted to assuage the Sultan's demands for reprisals against Yemen's support for PFLOAG.[159] (Border clashes between Oman and the PDRY in May of 1972 and October 1975 raised concerns in London about escalation.[160]) Interestingly, these operations, code named Operation Dhib, were, according to the defense scholar Geraint Hughes, contrary to the purposes of other cases of proxy warfare because "the UK's main purpose in supporting the Mahra's cross-border raids was to influence the policy of an ally (the Omani monarchy) rather than an adversary (the PDRY)."[161]

Phase III of the campaign lasted from 1973 to 1975. In 1973, the Iranians sent a brigade battle group because of the broader geopolitics of the region. (In 1968, the United Kingdom's Labour government of Harold Wilson, facing a devaluation crisis of the pound sterling declared that the British were going to leave the Gulf by 1971 and remove its commitments, except for Hong Kong, east of Suez.[162]) The Iranians wanted to ensure that they had a peaceful neighbor across the straits. The Jordanians also began to send in troops. The oil shocks of the early 1970s drastically increased Omani revenues and the Sultan began to spend it not only on development but also on military equipment. Government aid and development and support went from $60 million in 1971 to $1 billion in 1975.[163] (Oman's GDP went from $260 million in 1970 to $2.1 billion in 1975.[164])

The establishment of the defensive lines, the introduction of Iranian and Jordanian troops, and the expansion of SAF allowed for a continuation of population-centric COIN methods along with an increase in effective military operations to take place.[165] The scope of these operations and isolating the insurgents from their logistics and physical sanctuary across the border in the PDRY led to increasing numbers of surrendered enemy personnel and eventually to the insurgency petering out. (Between 1970 and 1974, for instance, 797 guerrillas defected to the government side.[166]) The Marxist insurgents started to have problems gaining support in a more traditional Muslim society.[167] In 1975, offensive operations by the Omanis and other forces broke the back of PFLOAG, which at this time had dropped the Occupied Arabian Gulf portion from its name and dispersed into more of a small-scale security problem.

CONCLUSIONS

This chapter provided cases illustrating various forms of indirect defensive interventions. The contention here is that while the United States may wish to move on from the COIN wars of the early twenty-first century, these modalities of conflict will still be a fact of life even in an era of GPC. When such modes of conflict need to be countered, the indirect approach is the most suitable way in dealing with them. As during the Cold War, great powers will still contest one another's aims both within their zones of influence and against those zones of their competitors. They will also conduct such competition in the strategic periphery. Similarly, nonstate actors, such as groups like al-Qaeda and the Islamic State, will continue to agitate across the Muslim world and other groups will rise, too. Middle powers, such as North Korea and Iran, will also seek to use unconventional capabilities abroad to provide defense in depth for their regimes.

The rise of the so-called gray zone conflict all but illustrates the necessity to be prepared for such defensive indirect interventions. Such conflicts have been defined as "the purposeful application of multiple elements of power—information, economic, military, political—to achieve objectives in ways that exceed the threshold for normal competition yet fall below the level of major interstate war."[168] Such conflicts are difficult because they tend to mitigate the possibility of wars between major powers, they hinder responses by "intentionally blurring peacetime and wartime operations," and they control escalation.[169] Responding to such events is especially difficult for a country like the United States because we tend to have difficulty understanding that the world does not operate in two separate spheres of "war" or "peace."[170]

What can we glean from the cases examined here?

The evidence presented earlier, in the main, suggests that U.S. service personnel at both the operational and tactical levels performed well and showed high political-military effectiveness in El Salvador between the years of 1981 and 1992. While the El Salvadoran Civil War ended in a tie, and cost the lives of approximately 75,000 of her citizens, democracy and political stability spread both there and in Central America. The stalemate and loss of Soviet support caused the FMLN to accept peace terms that were very favorable to the El Salvadoran government.[171] The costs to the United States—in terms of both combat fatalities and financial costs—were not exorbitant.[172]

American military advisors also showed a high degree of comfort and flexibility in operating in the small-war environment of El Salvador. The majority of the ESAF did eventually align itself with U.S. policy preferences (e.g., the marked drop in human rights abuses), the advisors showed initiative and adaptation in solving problems in a resource-constrained theater of operations, and autonomy was fully exercised. Hy Rothstein has argued strenuously that "El Salvador is an example where those closest to the action were inadvertently given sufficient flexibility to plan and operate in the context of benign neglect. Specifically, the focus of the senior US military brass remained fixed on Europe, regardless of the rhetorical importance of Central America for the Reagan administration."[173]

Of course, one might also conclude that too much autonomy led some advisors and military staff to be less than fully responsive to the policies and directives of the U.S. Country Team. The evidence here appears to be mixed. Yes, clearly some advisors played fast and loose with some directives in the country, but there also appeared to be some personality issues between certain ambassadors and elements of the military that perhaps amplified such incidents—especially as the tensions arose between the authorities of the SOUTHCOM GCCs and the COMs.

The Philippines and Trans-Sahara cases present a mixed bag in terms of the political-military effectiveness for post-9/11 light footprint, limited defensive indirect interventions advising missions. OEF-P definitely showed higher political-military effectiveness than in case of OEF-TS. In both areas of civil-military orientation, OEF-P showed higher levels of performance. The Country Team was at the forefront of operations and members of the

JSOTF-P showed a high degree of comfort and flexibility in operating in the environment. Of course, just as was the case with the 7th Special Forces Group (Airborne) in El Salvador, the fact that 1st Special Forces Group has had institutional connections in working with the AFP going back to 1957 perhaps gives it a leg up.[174] In addition, OEF-P personnel and staff only had to coordinate through one embassy[175] and the close geographical proximity of 1/1SFG(A) on Okinawa—akin to the close proximity of 3/7 SFG (A) to El Salvador—likely compounds that advantage, particularly as OEF-TS saw the rotation of different types of forces, and not just SOF personnel, through the area.[176] In addition, the endemic poverty and vast geographical expanses of political space comprising the Pan Sahel also are huge challenges—a comparison of tables 3.2 and 3.3, for instance, shows that after a decade of assistance manpower numbers remain the same, or in the case of Chad have decreased, while the numbers of ground vehicles and aircraft have not grown appreciably. The geographical realignment of the 3rd Special Forces Group (Airborne) back to the AFRICOM AOR might possibly, with time, help to increase its operational capabilities and effectiveness in the Sahel, but that is far from certain. The Trump administration had signaled that the U.S. military presence in Africa would shrink.[177]

Last, while Phase I of the Dhofar campaign was not a success, Phases II and III most definitely were. The campaign's success, however, was constituted on many factors. First, the removal of Said bin Taimur and his replacement with Qaboos bin Said allowed for a dramatic shift in government legitimacy. Sultan Qaboos's policies dramatically improved the situation on the ground. Second, while the seconded British personnel were ultimately responsible to the decisions of the Sultan, their involvement from the operational to tactical levels of decision-making made the situation unique. Third, the war was fought largely outside the spotlight of world events. The Omani government tightly controlled press access—although the BATT found it useful to provide warts and all radio-reporting of events on the ground there in order to counter communist propaganda from across the border in Yemen. Fourth, fighting a Marxist opponent in a traditional Muslim society allows for a certain asymmetry in the messaging once a more legitimate government comes to power—and the fact that the Sultan's mother was Dhofari also provided additional legitimacy to him. Fifth, the Dhofar war was not simply a battle between two distinct operational bins of either winning hearts and minds (COIN) or bold aggressive maneuver (conventional operations). It is not enough to simply point to the legitimate and multitude of heroic BATT tactical operations in the campaign and give them full credit for victory. Similarly, the conventional phase of operations at the close of the war would not have been possible without the development and preparation that Phase II of operations made possible.[178] Involvement by the Iranians and Jordanians also helped, but for an intervention of about 300 British personnel at any given time you had a very small-scale intervention that had arguably a strategic impact on regional geopolitics.

None of the above should be construed to argue that small-scale, long-duration FID campaigns will be the future operational norm. One can, however, look at the historical trends in how the U.S. military has been used abroad and make the assumption that such FID campaigns will be more likely than large-scale, long-duration COIN campaigns.

NOTES

1. See, for example, Max Boot, *The Road Not Taken: Edward Lansdale and the American Tragedy in Vietnam* (New York: Liveright, 2018).
2. *Joint Publication 3–22: Foreign Internal Defense* (Washington, DC: Joint Chiefs of Staff, August 17, 2018), ix.
3. While these programs take certain liberties with dialogue and the amalgamation of characters, they do show useful bits and pieces of FID. On *Narcos*, in particular, see my "'John Wayne Only Exists in Hollywood': *Narcos* and Foreign Internal Defense," *War on the Rocks* (January 13, 2016). Available at: https://warontherocks.com/2016/01/john-wayne-only-exists-in-hollywood-narcos-and-foreign-internal-defense/.
4. *Joint Publication 3–22: Foreign Internal Defense*, xi.
5. BASF's famous advertising jingle is: "At BASF, we don't make a lot of the products you buy. We make a lot of the products you buy better." See my "The Rise of the 'BASF' Doctrine?" *Geopoliticus: The FPRI Blog* (March 17, 2013). Available at: https://www.fpri.org/2013/03/the-rise-of-the-basf-doctrine/.
6. Lieutenant Colonel Brian Petit, "OEF-Philippines: Thinking COIN, Practicing FID," *Special Warfare* 23, no. 1 (January–February 2010): 12.
7. For an excellent discussion of covert versus clandestine operations, see Paul J. Tompkins Jr., Erin N. Hahn, and W. Sam Lauber, *Legal Implications of the Status of Persons in Resistance* (Ft. Bragg, NC: United States Army Special Operations Command, Assessing Revolutionary and Insurgent Strategies, no publication date), 120–129. Available at: https://www.soc.mil/ARIS/books/pdf/ARIS_Legal_Status-BOOK.pdf.
8. United States Army Special Operations Command, *Counter-Unconventional Warfare White Paper* (Ft. Bragg, NC: USASOC, September 26, 2014).
9. *Ibid.*, 4.
10. Thanks to Anna Simons for this insight.
11. Central Intelligence Agency, "El Salvador," *The CIA World Factbook* (online). Available at: https://www.cia.gov/library/publications/the-world-factbook/geos/es.html (accessed June 1, 2012).
12. See Richard A. Haggarty, ed., *El Salvador: A Country Study* (Washington, DC: GPO for the Library of Congress, 1988). Available at: http://countrystudies.us/el-salvador/ (accessed June 1, 2012).
13. As delivered by President Jimmy Carter, "Address at Commencement Exercises at the University," Notre Dame University, South Bend, IN, May 22, 1977. Available at: http://www.presidency.ucsb.edu/ws/index.php?pid=7552 (accessed December 12, 2012).

14. Today the title CINC is gone after Secretary of Defense Donald Rumsfeld replaced it with the moniker GCC due to his concerns over the civil-military relations implications of having multiple "Commanders-in-Chief."

15. Kalev Imants Sepp, "The Evolution of United States Military Strategy in Central America, 1979–1991" (PhD, Dissertation, Harvard University, February 2002), chapter 1.

16. *Ibid.*

17. AMB Robert E. White would only allow single augmentees to the MILGROUP for limited training. The ESAF sent soldiers to the School of Americas in Panama. About 300 attended a three-week-long "Human Rights Aspects in Internal Defense and Development" course in 1980. Later he allowed an MTT to train the ESAF on nonviolent crowd control techniques in anticipation of public demonstrations in January 1980. *Ibid.*, 28. Later, Ambassador Deane Hinton also had issues with LTG Nutting going so far as to declare, "I do not consider General Nutting as very politically astute." See his "Friction between Cincsouth and the Country Team," in *El Salvador at War: An Oral History of the Conflict from the 1979 Insurrection to the Present*, edited by Max G. Manwaring and Court Prisk (Washington, DC: National Defense University Press, 1988), 104.

18. The size of SF elements was decimated in the aftermath of Vietnam. See, for example, Susan Marquis, *Unconventional Warfare: Rebuilding U.S. Special Operations Forces* (Washington, DC: Brookings Institution Press, 1997). Conventional soldiers undoubtedly blamed the Special Forces in part for defeat in Vietnam. Bruce Hoffman states that all the material on COIN housed at the John F. Kennedy Special Warfare Center and School at Ft. Bragg, NC, was destroyed at the express direction of the senior army leadership after Vietnam. See his *An Agenda for Research on Terrorism and LIC in the 1990s*, RAND Paper, No.P7751 (Santa Monica, CA: RAND Corp., 1991), 2. Another thing that hampered U.S. political-military effectiveness early on in El Salvador was the decimation to active duty CA and Psychological Operations forces caused by post-Vietnam force reductions. But those units started to return in the 1980s and that helped reinforce human rights training at the School of the Americas and at ESAF training areas. Sepp, *The Evolution of United States Military Strategy*, 38.

19. *Ibid.*, 29.

20. *Ibid.*, 29–30. Retired major general Geoffrey Lambert, then a captain and the operations officer for 3/7SFG(A), later recalled in an interview: "First of all, considering that U.S. support would be reliant on the Salvadoran national will, and that advisory assistance was such a politically sensitive subject in the U.S. so soon after Vietnam, we shouldn't have proposed such a large package. Had we been sensitive to this, we probably could have proposed a 100-man package, and it probably would have done a lot more things. We had no idea that the 55-man limit would somehow become a ceiling, but it did. We were absolutely amazed that it got approved." "Interview: Special Forces in El Salvador," *Special Warfare* 6, no. 4 (October 1993): 35.

21. According to Sepp, however, no one from SOUTHCOM was involved with or knew of the covert and paramilitary campaign against the Sandinistas in Nicaragua in 1981. Sepp, *The Evolution of United States Military Strategy*, 19.

22. *Ibid.*, 35.

23. One OPATT officer, Lieutenant Colonel David Kindler, described this as "trying to coach a football game inside the locker room." *Ibid.*, 36–37.

24. His findings and suggested courses of action appeared as *Report of the El Salvador Military Strategy Assistance Team (Draft)* (San Salvador, El Salvador, September 12–November 8, 1981).

25. According to retired OPATT officer Cecil E. Bailey, "The Salvadoran Army [previously] had no experience, training or doctrine for counterinsurgency warfare. Much of the army's manpower was committed to static defense of critical sites and the economic infrastructure, forfeiting its offensive potential." "OPATT: The U.S. Army SF Advisers in El Salvador," *Special Warfare* 17, no. 2 (December 2004): 18.

26. *Ibid.*

27. Sepp, *The Evolution of United States Military Strategy*, 108. Retired SF Colonel, and El Salvador veteran, Hy Rothstein goes further still when he said:

This cadre predominately comprised Special Forces majors and master sergeants, although a few service members from other branches and services also became advisors. They all spoke Spanish and were culturally savvy. An informal Special Forces network debriefed outgoing advisors and prepared incoming advisors before their deployment to El Salvador. This informal system created a strategic coherence that was never evident in Washington, in the U.S. embassy or in the Regional Combatant Commander's headquarters.

Hy S. Rothstein, "Less Is More: The Problematic Future of Irregular Warfare in an Era of Collapsing States," *Third World Quarterly* 28, no. 2 (March 2007): 293.

28. "The way that the OPATTs succeeded was by going into the brigade and helping the brigade commander accomplish his agenda. I don't know that we could have imposed an outside solution, no matter how logical, integrated and good it was, because the brigade commanders wouldn't have accepted it. The program had to come from the ground up." "Interview: Special Forces in El Salvador," 36.

29. *Ibid.*, 33–34. From 1983 to 1991 the U.S. embassy reported the exact number of U.S. military personnel present each day in El Salvador by name. *Ibid.*, 109. See also U.S. General Accounting Office, "El Salvador: Extent of U.S. Military Personnel in Country," GAO/NSIAD-90–227FS, July 9, 1990.

30. Colonel James J. Steele, "The View from the Mil Group," in *El Salvador at War: An Oral History of the Conflict from the 1979 Insurrection to the Present*, edited by Max G. Manwaring and Court Prisk (Washington, DC: National Defense University Press, 1988), 407.

31. Bailey, "OPATT," 24.

32. Sepp, *The Evolution of United States Military Strategy*, 33–40. See also Christopher Dickey, "U.S. Relieves 3 of Salvadoran Duties," *Washington Post* (February 6, 1983), final edition. One of the soldiers relieved, Staff Sergeant Jay T. Stanley, was wounded from FMLN ground fire while directing ESAF ground combat operations from a helicopter.

33. Scott W. Moore, "Gold, Not Purple: Lessons from USAID-MILGP Cooperation in El Salvador, 1981–1992" (Master's Thesis, Naval Postgraduate School, December 1997), 34–35.

34. Quoted in *Ibid.*, 3.

35. *Ibid.*, 2.

36. "Interview: Special Forces in El Salvador," 37.

37. Chris L. Lukasevich, *U.S. Training and Advisory Assistance to the Armed Forces of El Salvador from 1981–1991 and the Resulting Decline in Human Rights Abuses: The Role of U.S. Army Special Forces* (MA Thesis, University of Arizona, 2002), 45. The ESAF grew from 11,000 conscripts to over 55,000 members during that time frame.

38. John D. Waghelstein, "Ruminations of a Pachyderm or What I Learned in the Counter-Insurgency Business," *Small Wars and Insurgencies* 86, no. 1 (Winter 1994): 364–365.

39. John D. Waghelstein, "Military-to-Military Contacts: Personal Observations—the El Salvador Case," *Low Intensity Conflict & Law Enforcement* 10, no. 2 (Summer 2001): 27.

40. *Ibid.*, 39.

41. "You have got to be able to do the analysis necessary to come up with some proposed courses of action. You know, there's nothing magic in this. The problem is that to arrive at the same conclusion, you have to understand what's pushing it. You can't look at El Salvador if you don't understand the *Matanza*. You can't look at El Salvador unless you understand the impact of the Soccer War. You can't look at El Salvador unless you understand the population pressures, land distribution, pre-revolutionary situation. You can't look at El Salvador unless you understand why they are the way they are, why the insurgents are doing what they're doing. In any situation like this you really have to do your homework first. Once you do that, from the military side you've got to understand the organization and why they're organized that way, what the historical precedences are, why they have a *tanda* system, or whatever system they have. What's the relationship between the officers and the troops? What do the troops do day to day? How responsive are they? How brave are the soldiers? With the right kind of training are they capable of taking on the guerrillas head on? What are the relationships of the population toward the military? Do they view the military as the reason for the problems or as a possible solution for the problem? You have to do your homework, do an analysis and really understand the nature of the beast you are going to be dealing with. And not be afraid to say what's right and what is wrong and if there's an impediment." Colonel John D. Waghelstein, "The United States Needs People Capable of Analyzing and Understanding a Problem Like the Salvadoran War," in *El Salvador at War: An Oral History of the Conflict from the 1979 Insurrection to the Present*, edited by Max G. Manwaring and Court Prisk (Washington, DC: National Defense University Press, 1988), 242.

42. According to Sepp, "The popular sense was that Fronius was a career sergeant, a professional soldier and a volunteer for hazardous service in El Salvador. His death in combat was not to be considered the same way at a nineteen-year-old private on his first enlistment, whose life was only to be placed at risk in defense of the United States against a clear and present danger. Fronius' death was a reminder of the risks for U.S. personnel working in El Salvador, but proved the value of the use of Special Forces personnel as instruments of the U.S. policy in political ambiguous situations." Sepp, *The Evolution of United States Military Strategy*, 122–123.

On the lack of large numbers of U.S. casualties in El Salvador, Lieutenant Colonel Pedrozo offered the following explanation: "I think an important point to make about Special Forces participation was the quality of the individual. The selection process for trainers was very rigorous. The trainers had to meet certain criteria, and I think that one of the reasons that we didn't have more people killed in El Salvador was the quality of the guy that we sent down there. The fact that we were able to do as much as we did is a good reflection on the training that we received in Special Forces." "Interview: Special Forces in El Salvador," 36.

43. "A careful review of available personnel rosters compiled by Colonel John McMullen and five other senior officers with El Salvador combat experience show that 291 CIBs/CMBs were earned." Gregory A. Walker, "Awards Honor SF Service During War in El Salvador," *Special Warfare* 11, no. 3 (Summer 1998): 38. Another announcement in the same issue stated, "[Special Forces], infantry and medical personnel who served in combat in El Salvador between Jan. 1, 1981, and Feb. 1, 1992, are eligible for the CIB and CMB." MAJ Tom McCollum, "USASOC Honors Veterans of El Salvador Combat," *Special Warfare* 11, no. 3 (Summer 1998): 46.

44. Robert D. Ramsey III, *Global War on Terrorism Occasional Paper 18: Advising Indigenous Forces: American Advisors in Korea, Vietnam, and El Salvador* (Fort Leavenworth, KS: Combat Studies Institute Press, 2006), 103.

45. Andrew J. Bacevich, James D. Hallums, Richard H. White, and Thomas F. Young, *American Military Policy in Small Wars: The Case of El Salvador* (New York: Pergamon-Brassey's, 1988). This report, however, may be faulted by perhaps being too influenced by the ongoing, slow-moving pace of the campaign at the time it was written.

46. "The other thing that I fought like hell for, and had an enormous problem with, was longer tours for military people. I felt that we were constantly running people through there who had to relearn. The one-year tour did not become effective for four to six months, and it was a tragedy that we did this. We didn't have that many people who wanted to come, first, and secondly, we didn't have that many people who could pick up as rapidly on what their predecessors had done, so in a sense we were constantly relearning old lessons. The notion that we could bring people out for longer and build into the system some institutional knowledge was another key to how we pushed the peanut ahead." Ambassador Thomas Pickering, "Continuity of Personnel Is Important," in *El Salvador at War: An Oral History of the Conflict from the 1979 Insurrection to the Present*, edited by Max G. Manwaring and Court Prisk (Washington, DC: National Defense University Press, 1988), 243–244.

47. *Ibid.*, 52–53.

48. Paul F. Gorman, "I.C. Eaker Distinguished Lecture on National Defense Policy," delivered at the U.S. Air Force Academy, Colorado Springs, CO, April 21, 1986. Sepp, *The Evolution of United States Military Strategy*, 58.

49. *Ibid.*, 79.

50. *Ibid.*, 90.

51. *Ibid.*, 147.

52. *Ibid.*, 149.

53. *Ibid.*, 192–194. This is unsurprising to the author. Having served alongside the Iraqi army during the darkest days of the Iraqi sectarian civil war, there were constant rumors and innuendo that senior officers were pocketing illicit funds from the sale of fuel vouchers, pocketing the earnings of soldiers that existed only on paper ("ghost soldiers"), and the sale of supplies and other equipment. War, insecurity, and military assistance served as the lubricant for HN—and sometimes U.S. service personnel and contractor—corruption.

54. *Ibid.*, 198.

55. "Interview: Special Forces in El Salvador," 38.

56. Central Intelligence Agency, "Philippines," *The World Factbook*. Available at: https://www.cia.gov/library/publications/the-world-factbook/geos/rp.html (accessed June 1, 2012).

57. Dr. C. H. Briscoe and Lieutenant Colonel Dennis Downey, "Multiple Insurgent Groups Complicate Philippine Security," *Special Warfare* 17, no. 1 (September 2004): 13.

58. PACOM as of 2018 is now called Indo-Pacific Command

59. Dr. C. H. Briscoe, "Balikatan Exercise Spearheaded ARSOF Operations in the Philippines," *Special Warfare* 17, no. 1 (September 2004): 16.

60. "Mission execution for this effort fell to the 1st Special Forces Group. Special Forces planners, based on extensive long term relationships with the Armed Forces of the Philippines (AFP), particularly with Philippine Special Forces and Scout Rangers, developed the detailed equipment requirements and a 6 month training program of instruction to provide a comprehensive counterterrorism capability for rural and urban situations. Training included urban and rural tactical training, assault of multiple structures, communications, intelligence, and medical training and sniper operations. This MTT was executed from January through July 2001 at Fort Magsaysay on the northern island of Luzon. As previously mentioned, on May 27, 2001, the Dos Palmas kidnapping occurred on Palawan island by members of the ASG who brought the hostages to Basilan Island at the southern tip of Mindanao. The Philippine military wanted to cease the LRC MTT and deploy the force to Basilan to support ongoing military operations to recover the hostages. However, after much discussion, the AFP decided to allow the training to be completed. Graduation was held on 6 July 2001 and the LRC deployed to Basilan on 8 July." Colonel (ret.) David S. Maxwell, *Foreign Internal Defense: An Indirect Approach to Counter-Insurgency/Counter Terrorism, Lessons from Operation Enduring Freedom- Philippines for dealing with Non-Existential Threats to the United States* (Written supplementary testimony presented to the House Armed Services Committee Subcommittee on Emerging Threats and Capabilities on Understanding Future Irregular Warfare Challenges, March 27, 2012), 6.

61. Colonel David S. Maxwell, U.S. Army, "Operation Enduring Freedom—Philippines: What Would Sun Tzu Say?" *Military Review* 84, no. 3 (May–June 2004): 20.

62. The SOCPAC Commander is the advisor for special operations on the PACOM staff. The staff is organized with a command group, 6 directorates (SOJ1-SOJ6), and is augmented by the Joint Intelligence Support Element/Joint Intelligence Center Pacific (JISE/JICPAC) and a theater signal detachment from the Army's 112th Signal Battalion (Special Operations) (Airborne).

> Part of SOCPAC's capability is based around Joint Task Force 510 (JTF-510), a rapidly deployable JTF Headquarters, which is nested within the command. JTF-510 provides the USPACOM commander with the ability to quickly establish command and control in support of emerging crises, such as disaster relief for tsunamis or earthquakes, humanitarian assistance for civil strife or non-combatant evacuation operations, or threat situations involving terrorist incidents.
>
> SOCPAC's strategy rests on a synchronized concept of operations called the Indirect Approach. The Indirect Approach focuses on 3 lines of operation: increasing partner nation security capacity, improving information gathering and sharing, and securing the support of the population. Specific tools used by SOCPAC in support of these lines of operations include the following: Joint and Combined Exchange

> Training (JCET); Counternarcotics Training (CNT); Foreign Internal Defense (FID); Subject Matter Expert Exchange (SMEE); Humanitarian Assistance/ Disaster Relief; Humanitarian Civic Action Programs; Humanitarian Mine Action (HMA); Information Operations and Public Affairs; Pacific Area Special Operations Conference (PASOC); and Joint Chiefs of Staff/PACOM Exercises.
>
> SOCPAC's assigned forces as of 2007 consisted of one Army Special Forces Battalion (1/1st Special Forces Group (Airborne), one Naval Special Warfare Unit (Naval Special Warfare Unit One), and one Air Force Special Operations Group (353rd Special Operations Group), which included 2 Air Force Special Operations Squadrons and one Air Force Special Tactics Squadron. Additionally, SOCPAC maintained a Joint Special Operations Air Component (JSOAC) in Hawaii and a forward-based Joint Special Operations Air Detachment (JSOAD) in the Philippines.

GlobalSecurity.org, "Special Operations Command, Pacific (SOCPAC)." Available at: http://www.globalsecurity.org/military/agency/dod/socpac.htm (accessed June 1, 2012).

63. Briscoe, "Balikatan Exercise Spearheaded ARSOF Operations in the Philippines," 17.

64. Maxwell, *Foreign Internal Defense*, 7.

65. *Ibid.*

66. Briscoe, "Balikatan Exercise Spearheaded ARSOF Operations in the Philippines," 18.

67. Maxwell, *Foreign Internal Defense*, 8.

68. *Ibid.*, 10–11. Army lieutenant colonel Brian Petit echoes this sentiment when he states as follows: "Contrary to popular perception, the U.S. mission in the Southern Philippines is not [COIN]. COIN is the mission of the GRP. The U.S. mission is FID in support of the GRP COIN campaign. This distinction is critical for two reasons: (1) The GRP, not the U.S., is directly responsible for combating insurgents, terrorists and lawless elements; (2) the U.S. role is to support a sovereign nation in both building the capacity of its armed and civil-security forces, and applying that capacity against violent extremists operating in undergoverned regions." See his "OEF-Philippines: Thinking COIN, Practicing FID," 12.

69. Dr. Cherilyn A. Walley, "Impact of the Semipermissive Environment on Force-Protection in Philippine Engagements," *Special Warfare* 17, no. 1 (September 2004): 39. For another in-depth discussion of this particular issue, see LTC Alfred E. Renzi Jr., "The Military Cooperation Group" (Master's Thesis, Naval Postgraduate School, December 2006), 55.

70. Maxwell, *Foreign Internal Defense*, 9.

71. Emphasis in original. *Ibid.*, 11. Maxwell credits then LTC Bill Medina who was a planner in the SOCPAC J5 in 2006 and later Chief of Staff, JSOTF-P, 2007 for this mission statement.

72. Former JSOTF-P Commander Colonel David Maxwell is quoted in, Major Stuart L. Farris, U.S. Army, *Joint Special Operations Task Force- Philippines* (Monograph, School of Advanced Military Studies, U.S. Army Command and General Staff College, AY2009), 32.

73. For those curious about the makeup of the JSOTF-P, Major Stuart L. Farris, U.S. Army, sketches the composition as follows:

> The JSOTF-P consists of members from all armed services as well as various U.S. government agencies. The Army contribution typically consists of personnel to augment

> the JSOTF-P staff, a reinforced Special Forces company, a Civil Affairs company, and a Military Information Support Team (MIST). The Navy provides personnel to augment the JSOTF-P staff and a Naval Special Warfare Task Unit (NSWTU) consisting of a SEAL platoon (+/-), a supporting boat detachment, and an Explosive Ordnance Disposal (EOD) detachment. The Air Force provides personnel to augment the JSOTF-P staff, liaison teams to coordinate with the Philippines Air Force (PAF), a weather detachment, and several fixed and rotary wing aircraft to support operations. Finally, general purpose forces (GPF), often in the form of U.S. Marines or Army infantry from the 25th Infantry Division or Guam National Guard, provide security for JSOTF-P base camps and facilities, and are also available to reinforce and support other operations when and where needed. Altogether, the combined number of U.S. service members assigned to the JSOTF-P is approximately 500.
>
> Staffed with less than seventy personnel, the JSOTF-P headquarters is located in the western archipelago town of Zamboanga on the island of Mindanao. The core operational elements of the JSOTF-P are the reinforced Army Special Forces Company, the NSWTU consisting of Navy SEALs, and the Army CA Company. Augmented by conventional forces and other supporting personnel (MISTs) when and where necessary, these core elements are task organized as cross-functional teams into three subordinate headquarters of approximately 100 to 150 personnel each and are known as Task Force (TF) SULU, TF, MINDANAO, and TF ARCHIPELAGO.

See his *Joint Special Operations Task Force-Philippines*, 38–39.

74. Colonel David Maxwell, quoted in Farris, *Joint Special Operations Task Force-Philippines*, stated:

> Over the years whenever there were security issues in the Philippine SOCPAC would deploy liaison elements (PSAT) to augment embassy military support. The presence of U.S. Special Operations personnel at country team meetings was not unusual. When the JSOTF was established the Ambassador and the Country Team welcomed the permanent JSOTF liaison elements that were an essential element for coordinating and approving myriad actions. These ranged from exchanging information on the situation in Mindanao, to approving psychological operations products that supported the State Department's Rewards for Justice program to coordinating development activities with USAID and media events with the Public Diplomacy officer. Although there was a permanent liaison presence from the JSOTF in the embassy to ensure transparency and synchronization, the JSOTF Commander, Commander Sergeant Major, Intelligence and Operations officers would fly to Manila to meet with the intelligence working group and attend both the Country Team and the political-military meeting. The commander and the Ambassador also met weekly to exchange information, provide the opportunity for the Ambassador to provide guidance and to synchronize activities with [the] Philippine government and military officers as well as media engagement. This close liaison and coordination was critical to ensure that all representatives of the U.S. government were speaking with one voice when dealing with counterparts and the press.

75. Maxwell, *Foreign Internal Defense*, 11–12. Army lieutenant colonel Brian Petit makes a similar argument when he stated:

> OEF-P is unique in that it was conceptualized and implemented by a small nucleus of Special Operations Forces. Special Operations Command Group (Airborne) implemented the "indirect approach" methodology, applying U.S. capacity strictly "through or with" the Armed Forces of the Philippines against the enemy and for the population. The indirect approach is both a philosophy and a method that is inculcated into all

> practitioners. The heart of the strategy is based on building relationships, reinforcing legitimate institutions, building security-force capabilities, sharing intelligence and information, developing focused civil-military programs, and aggressively promoting local acts of good governance. The indirect approach requires the discreet application of U.S. influence and assistance. Leaders continually calibrate the political implications of their actions, and quickly implement adjustments at the local level. The U.S. mission is led by the Joint Special Operations Task Force—Philippines.

See his "OEF-Philippines," 11.

76. For instance, Command Sergeant Major William Eckert noted that "the information gathered early on, combined with the overall plans of the AFP leadership, allowed the JSOTF-P to prepare focused subject-matter expert exchanges through which the AFP units acquire the skills needed to gain and maintain security within the joint operations area. During the time the JSOTF-P has focused on Sulu, the subject-matter-expert exchanges have been conducted with the AFP on an almost daily basis, including topics such as the combat-lifesaver course, small-unit tactics, marksmanship, maritime interdiction operations, radio communications, night-vision goggle use, close air support and leadership development." See his "Defeating the Idea: Unconventional Warfare in the Southern Philippines," *Special Warfare* 19, no. 6 (November–December 2006): 19.

77. In retelling an anecdote about building an outhouse in a village Major Herb Daniels relates: "Ironically, the reality of highly constrained resources encouraged ingenuity on the part of AFP/US personnel to develop a project which opened doors for continued military engagement. As the civil-military bonds strengthened, insurgent support diminished in a zero sum relationship—as the government got stronger, the insurgency got weaker. Identifying the needs of the population that could be satisfied by the AFP/US forces was the first critical step to destroying the links between the populace and the ASG and JI in Talipao. In the larger war on terror, thinking globally is important for policy makers, but acting locally will always be the greatest challenge for front line soldiers." See his *Keeping COIN Simple: The Outhouse Strategy for Security Development* (Hurlburt Field: The JSOU Press, 2009), 10.

78. Quoted in Max Boot and Richard Bennet, "Treading Softly in the Philippines: Why a Low-intensity Counterinsurgency Strategy Seems to Be Working There," *The Weekly Standard* (January 1, 2009).

79. "Vital language skills were also lacking on the teams. Among the 1st SF Group personnel who initially deployed to the Philippines, only two Soldiers were of Filipino descent. While their Tagalog language skills were useful with Chavacano, a combination of Tagalog and Spanish, they helped little with the Tausug and Yacan dialects that are common to Basilan and the southern Philippines." Briscoe, "Balikatan Exercise Spearheaded ARSOF Operations in the Philippines," 21.

80. While not nearly as restrictive as the rules for El Salvador, it took six months at the outset of operations to get approval for advisors to embed with companies rather than the original restriction of staying at the battalion or higher levels. Colonel David S. Maxwell, U.S. Army, "Operation Enduring Freedom—Philippines," 20–23.

81. Roel Landingin, "Philippines Agrees Peace Deal with Rebels," *Financial Times* (October 7, 2012). Available at: http://www.ft.com/cms/s/0/a06165b0-1044-11e2-a5f7-00144feabdc0.html#axzz2DKfWAbU8 (accessed November 25, 2012).

82. He goes on to say, "Big Oil, too, has lured the United States to the region: America already imports some 15 percent of its oil from West Africa—a figure expected to rise to 25 percent within a decade. Also of concern are the Chinese, who have been investing heavily in the region and whose influence the United States wants to limit." Kaplan, "America's African Rifles," *The Atlantic Monthly* 295, no. 3 (April 2005): 91.

83. Karl Wycoff, "Fighting Terrorism in Africa," Testimony before the House International Relations Committee. U.S. House Subcommittee on Africa, Washington, DC (April 1, 2004). Available at: http://commdocs.house.gov/committees/intlrel/hfa92870.000/hfa92870_0.HTM (accessed June 1, 2012).

84. For an in-depth discussion of the Tuareg rebellions in Mali, see Major Joachin Famakan Cissoko, Malian Armed Forces, "A Third-World Country Facing Indigenous Insurgencies: The Malian Touareg Insurgency" (Master's Thesis, U.S. Army Command and General Staff College, 2011–2012).

85. These Arab Afghans wore the garb of Afghanistan, brought back valuable combat experience, and were devoted to a strict interpretation of Islam. They also developed interpersonal networks with other like-minded individuals who were emboldened by the Soviet defeat. One of these networks, of course, was Osama bin Laden's al-Qaeda organization. See, for instance, Mohammed M. Hafez, "Jihad after Iraq: Lessons from the Arab Afghans," *Studies in Conflict & Terrorism* 32, no. 2 (January 2009): 73–94.

86. Raffi Khatchadourian, "Arming the 'Camel Corps'," *The Village Voice* 51, no. 5 (February 1–7, 2006): 32.

87. *Globalsecurity.org*, "Pan Sahel Initiative (PSI)." Available at: http://www.globalsecurity.org/military/ops/pan-sahel.htm (accessed June 1, 2012).

88. *Globalsecurity.org*, "Operation Enduring Freedom—Trans Sahara (OEF-TS)." Available at: http://www.globalsecurity.org/military/ops/oef-ts.htm (accessed June 1, 2012). It went on to state:

> In addition, SOCEUR was to develop an operational reach that enabled the ability to detect, exploit, deter, seize, defeat, or destroy targets throughout the area of responsibility. SOCEUR will accomplish this through two lines of operation. The first was the unconventional warfare line of operation (the main effort), an indirect approach to achieving the strategic goal by conducting a broad spectrum of military and paramilitary operations of longer duration, predominantly conducted through, with, or by indigenous and/or surrogate forces. The second was the deliberate engagement line of operation, a direct approach to achieving the strategic goal primarily through the time sensitive planning and targeting process to conduct short-duration strikes against terrorist targets or networks, crisis response to humanitarian disasters, or [Noncombatant Evacuation Operations (NEOs)] to protect US citizens and interests.

89. *Globalsecurity.org*, "Operation Enduring Freedom—Trans Sahara (OEF-TS)." Available at: http://www.globalsecurity.org/military/ops/oef-ts.htm (accessed June 1, 2012).

90. Raffi Khatchadourian, "War in the Greatest Desert," *The Village Voice* 51, no. 4 (January 25–31, 2006): 28.

91. Craig Whitlock, "Taking Terror Fight to N. Africa Leads U.S. to Unlikely Alliances," *The Washington Post* (October 28, 2006), final edition. According to Raffi

Khatchadourian one of the ways he received protections was by marrying into a tribe to bond alliances. "War in the Greatest Desert," 28. This appears to be a common tactics, techniques, and procedures (TTP) employed by AQAM. As a Nigerian army major, Moussa Salou Bartnou told Robert Kaplan, "I wish the Americans could see how the rebels settle in our border towns in the desert and marry local girls so that they become invisible, so that you don't know who you have to fight." "America's African Rifles," 94. AQI also used this TTP.

92. Khatchadourian, "War in the Greatest Desert," 28–30.

93. According to former EUCOM commander, and later President Barack Obama's first National Security Advisor, Marine Corps GEN James L. Jones:

> EUCOM established *Joint Task Force AZTEC SILENCE* under the Commander of the U.S. Sixth Fleet in December 2003 to counter transnational terrorism in the under-governed areas of Northern Africa and to build closer alliances with those governments. In support of this, U.S. Navy intelligence, surveillance and reconnaissance assets based in Sigonella, Sicily were used to collect and share information with partner nations and their militaries. This robust cooperative ISR effort was augmented by the release of intelligence collected by national assets.

Statement of GEN James L. Jones, USMC, Commander, United States European Command Before the Senate Armed Services Committee, Washington, DC (March 1, 2005), p. 28. Available at: http://www.armed-services.senate.gov/statemnt/2005/March/Jones%2003-01-05.pdf (accessed June 15, 2012).

94. Khatchadourian, "War in the Greatest Desert," 30.

95. According to an unidentified U.S. CT official, "U.S. intelligence on terrorists was provided to the Malian government and then was later found among Saifi's people, in the hands of the very militants the government was supposedly targeting." Khatchadourian, "Arming the 'Camel Corps,'" 32.

96. Kaplan, "America's African Rifles," 91.

97. Craig Smith, "U.S. Looks to Sahara as Front in Terror War Militants Recruit and Stock Up on Arms," *International Herald Tribune* (May 12, 2004), final edition.

98. GEN Charles F. Wald, "New Thinking at USEUCOM: The Phase Zero Campaign," *Joint Forces Quarterly* 43 (4th Quarter, 2006): 73.

99. See, for instance, Jeremy Keenan, "US Militarization in Africa," *Anthropology Today* 24, no. 5 (October 2008): 16–20.

100. Jim Garamone, "NATO, European Command Working in Africa," *American Forces Press Service* (October 28, 2005).

101. For in-depth background on AQIM, see Colonel Khalifa Hajji, Tunisian Air Force, "The Origins and Strategic Objectives of the al-Qaeda Organization in the Islamic Maghreb (AQIM)" (Master's Thesis, Naval Postgraduate School, December 2009).

102. Joseph Felter and Brian Fishman, *Al-Qaida's Foreign Fighters in Iraq: A First Look at the Sinjar Records* (West Point, NY: Harmony Project, Combating Terrorism Center at West Point, December 2007), 8. Libyans made up 112 of the foreign fighters (18.8 percent), Algerians accounted for 43 (7.2 percent), Moroccans 36 (6.1 percent), and approximately 25 Tunisians out of 700 records found in the Sinjar records. *Ibid.*

103. On the matter of foreign fighter recruitment, see Clinton Watts, "Foreign Fighters: How Are They Being Recruited? Two Imperfect Recruitment Models,"

Small Wars Journal (June 22, 2008). Available at: http://smallwarsjournal.com/jrnl/art/foreign-fighters-how-are-they-being-recruited (accessed July 1, 2012).

104. Dell L. Dailey and Jeffrey G. Webb, "U.S. Special Operations Command and the War on Terror," *Joint Forces Quarterly* 40 (1st Quarter 2006): 46.

105. See, for instance: "Travel Warning: Niger," Bureau of Consular Affairs, U.S. Department of State (April 6, 2012). Available at: http://travel.state.gov/travel/cis_pa_tw/tw/tw_5691.html (accessed June 1, 2012).

> The Department of State continues to warn U.S. citizens of the risks of travel to Niger, and urges extreme caution due to increased kidnapping threats against Westerners. This replaces the Travel Warning for Niger dated August 5, 2011 to update information on security concerns.
>
> Al-Qaida in the Lands of the Islamic Maghreb (AQIM), a terrorist group, continues its attempts to kidnap Westerners in Niger, including U.S. citizens, and has been successful in kidnapping Europeans in the region. On January 7, 2011, two French nationals were kidnapped in the capital city of Niamey. They were found dead less than 24 hours later following a rescue attempt by French and Nigerien military forces. In September 2010, seven people, including five French citizens, a Togolese national, and a Malagasy citizen, were kidnapped by AQIM from the northern mining town of Arlit. Four French citizens are still being held hostage by AQIM. In April 2010, a French citizen and his Algerian driver were kidnapped. The Algerian was freed. AQIM claimed to have killed the French citizen in retaliation for the July attempted rescue operation conducted by Mauritanian and French military forces. In November 2009, heavily armed individuals attempted to kidnap U.S. Embassy officials in Tahoua. Although there have been no kidnappings in Niger since January 2011, the Department of State Worldwide Caution dated January 24, 2012, reminds U.S. citizens to maintain a high level of vigilance and to take appropriate steps to increase their security awareness when traveling in the region.
>
> Because of the security threat, the U.S. Embassy in Niamey continues to restrict the travel of U.S. government employees and official visitors in the areas north of Niamey. The U.S. Embassy also continues to evaluate proposed travel and official and personal activities for employees, on a case-by-case basis, in consultation with Nigerien security authorities. Recently, the possibility of violence related to extremist and criminal groups filtering up from northern Nigeria has led the Nigerien government to recommend armed escorts for travel in far eastern Niger.
>
> As a result of safety and security concerns, some organizations, including foreign companies, NGOs, and private aid organizations, have temporarily suspended operations in Niger or withdrawn some family members and/or staff.
>
> Although the U.S. government places the highest priority on the safe recovery of kidnapped U.S. citizens, it is U.S. government policy not to make concessions to kidnappers.

106. About 60 percent of the cocaine going to Western Europe comes through transshipment in western Africa. The money involved is so large that one plane was burned in northern Mali after unloading as many as ten tons of cocaine rather than be reused. Scott Baldauf, "Air Al Qaeda," *The Christian Science Monitor* (January 15, 2010), final edition.

107. "To finance their operations, the militants exact tolls from smugglers whose routes traverse the Qaeda sanctuary, and collect ransoms in kidnappings. In late October, two Austrians were released after a ransom of more than $2 million was

reportedly paid. They had been held in northern Mali after being seized in southern Tunisia in February." Eric Schmitt, "U.S. Helps African States Fend Off Militants," *The New York Times* (December 13, 2008), final edition.

108. Hoffman has used this term in numerous conversations with the author post-2007.

109. See, for instance, Frank G. Hoffman, "Hybrid Threats: Reconceptualizing the Evolving Character of Modern Conflict," *Strategic Forum No. 240* (April 2009). Available at: http://www.ndu.edu/inss/docuploaded/StrategicForum_240.pdf (accessed June 1, 2012).

110. William Maclean, "Local Wars Blur Al Qaeda's Threat to West," Reuters.com (July 5, 2012). Available at: http://www.reuters.com/article/2012/07/05/us-security-qaeda-idUSBRE86408D20120705 (accessed July 15, 2012). The article also states that in June 2012 GEN Carter Ham, the former AFRICOM commander, said that "Boko Haram [a Nigerian violent extremist group], al Shabaab and AQIM were sharing funds and swapping explosives in what could mark a rise in the security threat on the continent." See also Lauren French, "African Extremist Groups Linking Up: U.S. General," *Reuters.com* (June 25, 2012). Available at: http://www.reuters.com/article/2012/06/25/us-security-usa-africa-idUSBRE85O18P20120625 (accessed July 15, 2012).

111. *U.S. Africa Command*, "Fact Sheet: United States Africa Command." Available at: http://www.africom.mil/getArticle.asp?art=1644 (accessed June 1, 2012).

112. *U.S. Africa Command*, "Fact Sheet: Trans-Sahara Counterterrorism Partnership." Available at: http://www.africom.mil/tsctp.asp (accessed June 1, 2012). The fact sheet goes on to say:

> U.S Africa Command facilitates programs that expand military-to-military cooperation between the US and partner nations by ensuring there is adequate resources available to train, advise, and assist regional units with the hope of promoting better regional cooperation, communications, and intelligence sharing. U.S. Africa Command will conduct train and equip missions to build Company level units that are capable of providing sustained border security. Also U.S. Africa Command facilitates partner national staff information training and mentorship programs, conduct [Joint Planning Assistance Teams] and [Military Information Support Teams] training, and hold [Joint Combined Exercise Training] events in all nine countries. To build regional cooperation, an annual regional [CT] exercise will involve all countries as well as mentors and observers from African partners. As the military support to TSCTP, OEF-TS is currently building an information sharing portal and standing up regional communications infrastructure to improve connectivity between partner countries.

113. According to an analysis conducted by Lianne Kennedy Boudali:

> The program budget is made up of a combination of funds requested by the Department of State, USAID, and the Department of Defense. The TSCTP has been supplemented by "Section 1206" and "Section 1207" funds, which provide the Secretaries of State and Defense with spending authority to jointly implement programs to build capacity in partner nations by providing training and equipment to foreign military forces. The majority of the TSCTP budget will be spent on development initiatives such as improving health and education, building community centers, providing vocational training, supporting girls' scholarship funds, promoting transparency in governance, and providing support and training for community radio stations. Embassy teams work with EUCOM and USAID personnel to determine the package

> of development programs that will match the goals of the TSCTP to the needs of the host country.

See her *The North Africa Project: The Trans-Sahara Counterterrorism Partnership* (West Point, NY: Combating Terrorism Center, U.S. Military Academy, April 2007), 4–5.

114. Johnnie Carson, assistant secretary of state, Bureau of African Affairs, Opening Remarks for Hearing on Counterterrorism in Africa (Sahel Region), Senate Committee on Foreign Relations, Subcommittee on Africa, Washington, DC (November 17, 2009). Available at: http://www.state.gov/p/af/rls/rm/2009/132062.htm (accessed June 1, 2012).

115. F. John Bray, U.S. Foreign Service, *The Trans-Sahara Counterterrorism Partnership: Strategy and Institutional Friction* (Department of Military Strategy, Planning, and Operations, U.S. Army War College, 2011).

116. *Ibid.*, 12.

117. *Ibid.*, 12–13. Bray notes that he "was the newly arrived security officer [in Mauritania], who was also the back-up consular officer, the security assistance officer for the restarted IMET program, and on occasion the acting deputy chief of mission and charged 'affairs." *Ibid.*, FN 67, 28.

> Compounding this problem, . . . because of staffing demands for Afghanistan (2002) and Iraq (2003), hardship posts commonly faced severe staffing shortfalls in terms of personnel and talent. In 2006, GAO reported that mid-level positions at hardship posts were commonly staffed by junior officers, often on their first assignments. Moreover, there were significant deficiencies in the ability, training, and experience of officers serving in "stretch" assignments. Many junior officers stretched into more senior positions lacked managerial experience and the supervisory guidance to effectively do their jobs. As a consequence, senior staff, including ambassadors, spent more time on operational matters and less time on policy planning and coordination. As of 2006, many officers at hardship posts also lacked language proficiency for their positions: 77 percent in Mali; 20 percent in Mauritania; 57 percent in Chad; and 10 percent in Niger, did not meet specified language requirements.
>
> On 9/11 most of the country teams in Sahel barely had the capacity to support themselves, much less develop capacity for the host nation. Gaps in experience, skills and language proficiency inhibited their capability to accomplish their fundamental mission of Diplomacy, much less the Information, Military, Economic, Financial, Intelligence and Law enforcement functions (DIME-FIL) required of by a whole-of-government approach. Over time, things improved. For example, in Nouakchott during 2002 a general services officer was assigned to assist the post management officer and the CIA re-opened their station. In 2003 the defense attaché office reopened. A public diplomacy position and a political officer position were added in 2005, and a six-person Marine Security Guard detachment was added shortly thereafter. In 2006 a USAID coordinator was added to manage development programs, and there are plans to add a security assistance officer in 2011 to manage military-to-military cooperation programs. The net effect of the lack of capacity is that the DOS-managed embassy platform was not staffed to match the pace of USEUCOM's (and later USAFRICOM's) staffing and coordination initiatives. This leads directly to a third conclusion that can be drawn from the TSCTP: Although an Irregular Warfare strategy may be implemented at the embassy level, overseas diplomatic missions

> are not always staffed to meet the requirements of a whole-of-government approach; they will grow to meet the need, but not overnight."

Ibid., 13–14. The GAO report mentioned is *Staffing and Foreign Language Shortfalls Persist Despite Initiatives to Address Gaps* (Washington, DC: U.S. Government Accounting Office, GAO-06–894, August 2006), 19.

118. *Ibid.*, 15–18.

119. *Ibid.*, 20. On the preceding pages he notes:

> Although State, USAID, and the DoD jointly collaborated in implementing the TSCTP, the GAO found that disagreements about whether the ambassador should have authority over DoD personnel temporarily assigned to conduct TSCTP activities has hampered implementation. As evidence of this the report, DoD officials complained in 2007 when the ambassador in Niger limited the number of DoD personnel in country because of concerns over the country's fragile political environment and the limited support capacity in the embassy. In Chad, another ambassador called for a "strategic pause" to TSCTP activities to reassess the mission's ability to support activities. The authorities of the GCC are derived from Title 10, USC Section 164, under which the GCC is responsible for all military personnel and activities under their command. Chief of Mission authorities are outlined in Title 22 USC, 3902, 3927 and 4802, which specify Chief of Mission responsibilities for all executive agencies overseas except those deployed under the authority of a GCC. In the State Department's reply to the GAO, the friction is clearly articulated.

Ibid., 18–19.

120. For an example of one such training mission where teams from the 10SFG(A) and 20SFG(A) "lacked Africa expertise and were short on French and Arabic speakers," see Ann Scott Tyson, "U.S. Pushes Anti-Terrorism in Africa Under Long-Term Program, Pentagon to Train Soldiers of 9 Nations," *Washington Post* (July 26, 2005), final edition. See also David White, "Niger Enrolls in US Global War on Terror: Former French Colony Is Assuming Strategic Importance in Fight against Al-Qaeda and Its Allies in Africa's Sahel," *Financial Times* (October 1, 2004), final edition.

121. The JSOTF-TS was transferred to AFRICOM from EUCOM in October 2008 and then subsequently was named a subordinate unit of SOCAFRICA in May 2009. *GlobalSecurity.org*, "Joint Special Operations Task Force—Trans Sahara (JSOTF-TS)." Available at: http://www.globalsecurity.org/military/agency/dod/jsotf-ts.htm (accessed June 1, 2012).

122. See, for example, Major Danford W. Bryant II, "Into Africa: CA Teams Expand Operation Enduring Freedom into Africa," 18–25; and, MAJ John P. Wishart, "Out of Africa," *Special Warfare* 24, no. 4 (October–December 2011): 10–15.

123. See, for example, Stephen Ellis, "Briefing: The Pan Sahel Initiative," *African Affairs* 103, no. 412 (July 2004): 464.

124. For such a representative piece, see Vijay Prashad, "Fighting a Proxy War for America?" *New African* 490 (December 2009): 46–47.

125. See, for instance, International Crisis Group, *Islamist Terrorism in the Sahel: Fact or Fiction?* (Africa Report N°92–31 March 2005), 30; and, Joshua Keating, "Was the Mali Coup Leader Trained in the U.S.?" Available at; http://passport.foreignpolicy.com/posts/2012/03/23/was_the_mali_coup_leader_trained_in_the_us (accessed June 1, 2012).

126. International Crisis Group, *Islamist Terrorism in the Sahel*, 6.

127. Robert D. Kaplan, *Hog Pilots, Blue Water Grunts: The American Military in the Air, at Sea, and on the Ground* (New York: Random House, 2007), 304–305.

128. Dan Murphy, "Did Libya's Revolution Topple Mali into Crisis?" *Christian Science Monitor* (April 6, 2012). Available at: http://www.csmonitor.com/World/Backchannels/2012/0406/Did-Libya-s-revolution-topple-Mali-into-crisis (accessed November 15, 2012).

129. Craig Whitlock, "Questions Surround U.S. Deaths in Mali: Officials Say Little about 3 Commandos, 3 Women Killed in April Car Crash," *Washington Post* (July 9, 2012), final edition.

130. *Ibid.*

131. Adam Entous and Drew Hinshaw, "U.S. Sets Sights on Al Qaeda in Mali: Pentagon Weighs Action as Militants Gain Sway in Nation's Vast North," *Wall Street Journal* (July 27, 2012), final edition. Operating in ungoverned space might mean conducting the mission of UW. UW is defined as "activities conducted to enable a resistance movement or insurgency to coerce, disrupt, or overthrow a government or occupying power by operating through or with an underground, auxiliary, and guerrilla force *in a denied area*." [Emphasis added.] Department of Defense, "unconventional warfare," *DOD Dictionary of Military Terms*. Available at: http://www.dtic.mil/doctrine/dod_dictionary/ (accessed August 1, 2012). UW is the flipside of FID and SF are specially trained for this mission as they demonstrated at the outset of OEF and in northern Iraq at the outset of OIF.

132. See, for instance, Associated Press, "Rebels Destroy Historic Sites in Mali," *Wall Street Journal* (July 1, 2012), final edition; Entous and Hinshaw, "U.S. Sets Sights on Al Qaeda in Mali," final edition; and, Emily Alpert, "U.N.: Over 435,000 Displaced, Children Forced to Fight in Mali," *World Now* (August 17, 2012).

133. "AFRICOM chief Gen. Carter Ham says some of Benghazi attackers had al Qaeda 'linkages,'" *CBS News* (November 14, 2012). Available at: http://www.cbsnews.com/8301-202_162-57549657/africom-chief-gen-carter-ham-says-some-of-benghazi-attackers-had-al-qaeda-linkages/ (accessed November 25, 2012).

134. Helene Cooper, Thomas Gibbons-Neff, and Eric Schmitt, "Soldier in Bloody Niger Mission Had Warned of Gaps, Defense Officials Say," *The New York Times* (March 19, 2018).

135. Charlie Savage, Eric Schmitt, and Thomas Gibbons-Neff, "U.S. Kept Silent about Its Role in Another Firefight in Niger," *The New York Times* (March 14, 2018).

136. Thomas Gibbons-Neff, and Eric Schmitt, "Pentagon May Cut Commando Forces in Africa in Major Military Review," *The New York Times* (June 4, 2018).

137. Joe Penney, Eric Schmitt, Rukmini Callimachi, and Christoph Koettl, "C.I.A. Drone Mission, Curtailed by Obama, Is Expanded in Africa Under Trump," *The New York Times* (September 9, 2018).

138. "The Risks of Reducing U.S. Special Operations in Africa," *Council on Foreign Relations* (September 13, 2018). Available at: https://www.cfr.org/interview/risks-reducing-us-special-operations-africa.

139. See, for instance, J. E. Petersen, *Oman's Insurgencies: The Sultanate's Struggle for Supremacy* (London: Saqi Books, 2008); Major General Tony Jeapes, *SAS Secret War: Operation Storm in the Middle East* (London: Greenhill Books, 1980; 2005); and John Akehurst, *We Won a War: The Campaign in Oman 1965–1975* (Wilton, UK: M. Russell, 1982).

140. "Brief on Muscat and Oman," HQ, SAF, Bait-al-Falaj, Muscat (June 30, 1965), p. 4.

141. Ian Cobain, "Britain's Secret Wars," *The Guardian* (September 8, 2016). Available at: https://www.theguardian.com/uk-news/2016/sep/08/britains-secret-wars-oman?CMP=share_btn_tw.

142. See, for instance, Abdel Razzaq Takriti, *Monsoon Revolution: Republicans, Sultans, and Empires in Oman, 1965–1976* (New York: Oxford University Press, 2013).

143. "Brief on Muscat & Oman," p. 4.

144. Secondment is the temporary assignment of personnel from one organization to another.

145. For some background on this assistance and training, see, for example, Marc R. DeVore, "A More Complex and Conventional Victory: Revisiting the Dhofar Counterinsurgency, 1963–1975," *Small Wars & Insurgencies* 23, no. 1 (January 2012): 144–173.

146. Walter C. Ladwig III, "Supporting Allies in Counterinsurgency: Britain and the Dhofar Rebellion," *Small Wars & Insurgencies* 19, no. 1 (March 2008): 70.

147. Jeapes, *SAS Secret War*.

148. This Jebel Akhdar campaign occurred after the Suez Crisis of 1956. The British wanted to lend aid to the Sultan without using a large footprint. Two squadrons of SAS troops, roughly 140 men, on their way back from Malaya conducted daring operations that put down the rebellion. The success of this campaign, it is said, led to maintaining the SAS in the post-war period. See, for instance, Alastair MacKenzie, *Special Force: The Untold Story of 22nd Special Air Service Regiment (SAS)* (New York: I. B. Tauris, 2011), 81–93.

149. Jeapes, *SAS Secret War*.

150. *Ibid.*

151. Ladwig, *Supporting Allies in Counterinsurgency: Britain and the Dhofar Rebellion*, 71.

152. *Ibid.*, 79.

153. *Ibid.*, 74–75.

154. S. Monick, "Victory in Hades: The Forgotten Wars of the Oman 1957–1959 and 1970–76," *Scientia Militaria, South African Journal of Military Studies* 12, no. 4 (1982): 16.

155. Ian Illych Martinez, "The Battle of Mirbat: Turning Point in the Omani Dhofar Rebellion," *Small Wars & Insurgencies* 23, no. 3 (June 2012): 517–526.

156. Oral History 11,161, 1992–20–23, Imperial War Museum, London, United Kingdom (accessed on June 10, 2015).

157. "Muscat and Oman: Half Sovereign?" *The Economist* (October 18, 1969).

158. Geraint Hughes, "A 'Model Campaign' Reappraised: The Counter-Insurgency War in Dhofar, Oman, 1965–1975," *Journal of Strategic Studies* 32, no. 2 (April 2009): 294.

159. *Ibid.*, 295.

160. Geraint Hughes, "A Proxy War in Arabia: The Dhofar Insurgency and Cross-Border Raids into South Yemen," *Middle East Journal* 69, no. 1 (Winter 2015): 94.

161. *Ibid.*, 102.

162. Cobain, *Britain's Secret Wars*.

163. Hughes, *A 'Model Campaign' Reappraised*, 290.

164. Available at: https://www.macrotrends.net/countries/OMN/oman/gdp-gross-domestic-product.

165. Clive Jones, "Military Intelligence, Tribes, and Britain's War in Dhofar, 1970–1976," *The Middle East Journal* 65, no. 4 (Autumn 2011): 565.
166. Hughes, *A 'Model Campaign' Reappraised*, 283.
167. Jones, *Military Intelligence, Tribes, and Britain's War in Dhofar, 1970–1976*.
168. National Intelligence Council, "Foreign Approaches to Gray Zone Conflicts," March 2016. Available at: https://nsiteam.com/foreign-approaches-to-gray-zone-conflicts/.
169. *Ibid.*
170. On this see, Adam Elkus and Michael P. Noonan, "Competitive Shaping in World Politics: A Bibliographic Essay and Course Outline," *The Philadelphia Papers* #18 (Philadelphia, PA: Foreign Policy Research Institute, May 2018).
171. Conversation between the author and noted Latin America specialist Dr. John T. Fishel on September 18, 2008, at Ft. Leavenworth, KS.
172. Twenty-one U.S. Service personnel were killed in action in El Salvador compared to 3,540 KIA in Iraq or 1,897 KIA in Afghanistan as of January 21, 2021. To be clear, to the friends and families of these men and women who made the ultimate sacrifice even one KIA is too many, but here we are discussing vast differences in scale. Casualty data are taken from Tom Bowman, "Medal OK'd for Soldiers Who Served in El Salvador Army Reverses Decision Denying Combat Badge to Green Beret Advisers," *Baltimore Sun* (February 6, 1998), final edition; available at: http://www.defense.gov/news/casualty.pdf (accessed November 25, 2012).
173. Rothstein, *Less Is More*, 281. He then goes on to say, on the same page, that "in retrospect, these deficiencies created the possibility for a small cadre of dedicated and knowledgeable Special Forces soldiers to achieve US policy objectives. In Washington it seemed that the biggest concern regarding military involvement was not to exceed the 55-man ceiling. Politicians also demanded that military personnel function as trainers and not combat advisors. Minimizing military involvement in a communist-supported insurgency helped ease any similarity with Vietnam. The effect of Washington's benign neglect was to enable a protracted effort in El Salvador, which would not have been possible if thousands of US troops had been deployed in a direct combat role."
174. "The History of the 1st SF Group in the Republic of the Philippines: 1957–2002," *Special Warfare* 15, no. 2 (June 2002): 14–15.
175. Colonel Gregory Wilson reiterates the importance of working with the Country Team when he wrote as follows:

> In countries with well-established governments, WOT military operations play a supporting role to efforts managed by the U.S. State Department. Planning that integrates the military and country-team staff members produces optimal results. Because of the protracted nature of these operations, military and country-team staff must maintain close relationships and conduct interagency coordination on a regular basis. In the Philippines, OEF-P planners coordinate closely with the country team to facilitate interagency planning and synchronization.

Colonel Gregory Wilson, U.S. Army, "Anatomy of a Successful COIN Operation: OEF-PHILIPPINES and the Indirect Approach," *Military Review* 86, no. 6 (November–December 2006): 42.

176. In Mali, for instance, Special Forces major Simon J. Powelson offers chapter and verse for the limits of episodic engagement there with different forces moving

in and out, but not providing enduring engagement until in 2010 the 10th SFG(A) established a Special Operations Forces Liaison Element and a Joint Planning Assistance Team to track training and logistics to ensure that Malian units were building effective capabilities. See his "Enduring Engagement Yes, Episodic Engagement No: Lessons for SOF from Mali" (Master's Thesis, Naval Postgraduate School, December 2012). Available at: https://calhoun.nps.edu/handle/10945/38996.

177. Helene Cooper, Thomas Gibbons-Neff, Charlie Savage, and Eric Schmitt, "Pentagon Eyes Africa Drawdown as First Step in Global Troop Shift," *The New York Times* (December 24, 2019).

178. See, for example, Marc R. DeVore, "A More Complex and Conventional Victory: Revisiting the Dhofar Counterinsurgency, 1963–1975," *Small Wars & Insurgencies* 23, no. 1 (March 2012): 144–173; Geraint Hughes, "A 'Model Campaign' Reappraised: The Counter-Insurgency War in Dhofar, Oman, 1965–1975," *The Journal of Strategic Studies* 32, no. 2 (April 2009): 271–305; Thomas Mockaitis, *British Counter-Insurgency in the Post-Imperial Era* (Manchester: Manchester University Press, 1995); and, John Newsinger, *British Counterinsurgency: From Palestine to Northern Ireland* (Basingstoke, UK: Palgrave Macmillan 2002).

Chapter 4

Offensive Indirect Interventions—Unconventional Warfare

The United States and other actors also conduct offensive indirect interventions. These are referred to as UW in U.S. military doctrine. For the United States, SF are the principal military tool for such operations but are assisted by other units as well. The CIA is oftentimes a lead actor, if not the actor, for such operations. UW is used to overthrow a foreign regime or assist other actors on the ground. This chapter explores the U.S. use of such tools through the limited interventions of Angola (1975–1990), Nicaragua (1981–1991), and northern Iraq (1992–2002), and the assertive cases of Afghanistan (2001–2002) and northern Iraq (2002–2003).

States in the international system, whether great or small, will sometimes find it in their interests to militarily intervene in an offensive manner in the affairs of other states—this may occur across an entire country or within smaller portions of it. This may occur in more formalized uses of interstate violence, it may take place through the use of more discrete military or paramilitary forces conducting operations and/or supporting local antigovernment factions, or it can carry out the training and equipping of groups or elements within that country either within the boundaries of that state or in a neighboring sanctuary. Sometimes all of these elements can occur in either a staggered way or near-simultaneously. Such interventions may take place against a hostile regime or to support elements fighting against a third-party occupation. Depending on the objectives of the actors conducting such operations, the goal might be to remove a government from power or to alter conditions on the ground in the targeted state so as to change its domestic or foreign and defense policies that are seen as problematic by the intervening power(s).

INDIRECT INTERVENTIONS AND UNCONVENTIONAL WARFARE

As stated earlier in this book, in the U.S. military the support of such substate foreign elements is referred to as UW. The DOD defines UW as "activities conducted to enable a resistance movement or insurgency to coerce, disrupt, or overthrow a government or occupying power by operating through or with an underground, auxiliary, and guerrilla force in a denied area."[1] Within this definition, and in simplest terms, undergrounds refer to the intelligence networks of the resistance movement or insurgency,[2] auxiliaries are the logistics and support elements of those groups, and the guerrilla force are their fighting forces. But the U.S. military are not the sole actors in such interventions. The U.S. IC and other actors may play a role and even nongovernmental groups or individuals may be contracted to carry out various functions.

Retired Army colonel Abigail Linnington notes that, in UW, "SOF has focused more of its efforts on the organization and training of the guerrilla forces while the CIA has often led USG efforts on subversion, sabotage, and intelligence collection in the development of the underground."[3] And although SOF may be at the tip of the military spear in a UW campaign, and in particular SF, other non-SOF forces may also be involved. On the CIA side of things, members of the Directorate of Operations' Special Activities Division (SAD) and other elements are the key players. Subversion refers to "actions designed to undermine the military, economic, psychological, or political strength or morale of a governing authority."[4] This division of labor certainly makes sense in terms of authorities and oversight, but it can lead to trouble on the ground.

When UW is conducted below the threshold of interstate war, there is obviously much secrecy surrounding such activities. The opaque nature of this type of activity makes a robust analysis of such operations difficult and will necessarily make this chapter shorter than the previous one. In such environments, UW may be carried out as covert or clandestine operations. Covert operations are "so planned and executed as to conceal the identity of or permit plausible denial by the sponsor,"[5] whereas clandestine operations are "sponsored or conducted by governmental departments or agencies in such a way as to assure secrecy or concealment."[6] Covert actions are codified within the U.S. Code, while clandestine operations are not. In essence, covert actions are the strategic concealment of such interventions while clandestine operations are more concerned with tactical secrecy. Covert actions require a presidential finding. Historically, the CIA has had the lead in covert operations, but the president may designate other departments, agencies, or entities of the USG to conduct them.[7]

When UW operations are part of a more formalized use of force against another state or nonstate actor in a denied area, they usually take the form of low-visibility operations. Such activities are defined as follows:

> Sensitive operations wherein the diplomatic-military restrictions inherent in covert and clandestine operations are either not necessary or not feasible; actions are taken as required to limit exposure of those involved and/or their activities and with the knowledge that the action and/or sponsorship of the operation may preclude plausible denial by the initiating power.[8]

The early UW activities by the U.S. in Afghanistan and northern Iraq provide examples of such operations. U.S. CT operations in places such as Pakistan, Yemen, Somalia, and Libya also demonstrate these types of missions. The United States does not actively broadcast its presence, but operations on the ground, particularly of the precise and lethal variety, make it clear to enemy forces that they are being engaged by nonlocal actors.

Just as in the discussion of small footprint FID from the previous chapter, here UW is split between the categories of limited (i.e., the use of U.S. forces in combat operations is restricted) or assertive (i.e., U.S. combat involvement is relatively unrestricted). Once again, the "through, with, and by" construct might all take place in a limited offensive intervention, whereas an assertive offensive intervention would also entail a very robust form of "with." Table 4.1 offers some examples of each category.

Offensive indirect operations are carried out by a great power of global reach such as the United States for many of the same reasons defensive indirect operations are carried out: the capabilities and resources of the USG are finite. Even in an age of a disordered world, the United States must pick and choose where it can distribute and apply its elements of national power and weigh their use, hopefully, within a strategic framework that considers the where, when, how, and why of interventions. Absent such a framework, the United States is likely to fritter away its attention and resources and cause unintended second- and third-order effects.

There is a long history of such offensive indirect interventions conducted by both the United States and other actors. For example, during the American War for Independence, Francis Marion, the "Swamp Fox," conducted operations in British-occupied South Carolina. During the nineteenth century private U.S. citizens known as Filibusters tried to foment revolutions abroad. These

Table 4.1. Forms of Offensive Indirect Interventions

Limited	Angola (1975–1990), Nicaragua (1981–1991), Iraq (1991–2002)
Assertive	Afghanistan (2001–2002), northern Iraq (2002–2003)

"small groups of like-minded citizens would support revolutions abroad and in frontier territories where European powers still held de jure sovereignty."[9] The twentieth century, as discussed earlier, showed the rise of the OSS to conduct partisan operations behind enemy lines in Europe and Asia and then the global struggle between the United States and Soviet Union where many allies and client states were backed in wars for control in the periphery.

While some thought that the end of the Cold War would limit the scope of such competitions, they may be seen as the preferred form of conflict in the twenty-first century. The absolute scale and scope of such proxy wars may be smaller than during the geopolitical struggle of the Cold War, but powers large and small (but particularly large) view such conflict as a useful tool of statecraft. The current situation in Syria illustrates this perfectly: Iranian advisors, Russian forces and mercenaries, Hezbollah forces, and Shi'a Iraqi militias fight for the Alawite regime against foreign fighter volunteers from fifty-odd nations, with indirect military and nonmilitary assistance from the Gulf States, Turkey, and the United States, among others. Motivations for such support range from the sectarian to the strategic.[10]

But what are proxy wars? The British scholar Andrew Mumford defines such wars as

> the product of a relationship between a benefactor, who is a state or non-state actor external to the dynamic of an existing conflict, and the chosen proxies who are the conduit for the benefactor's weapons, training and funding. In short, proxy wars are the logical replacement for states seeking to further their own strategic goals yet at the same time avoid engaging in direct, costly and bloody warfare.[11]

They are seen as lower cost (in terms of blood and treasure) and offer plausible deniability. He explains that there are four ways such proxy wars can be conducted: (1) "a state uses another state (as a surrogate force)"; (2) "a state uses a non-state actor"; (3) "a non-state actor uses a state"; and (4) "a non-state actor uses another non-state actor (as a surrogate force)."[12] The United Sates has engaged in both type 1 and type 2 proxy wars. This chapter focuses on type 2.

Before we turn to the U.S. cases, however, it is important to note, as mentioned in chapter 1, how other states are conducting both limited and assertive indirect interventions. Russia has shown a particular skill in using such method in Georgia (2008), Crimea (2014), and eastern Ukraine (2015–present). Iran has also shown skill in fomenting unrest in Iraq (2003–present), Gaza (1990–present), Yemen (mid-2000s–present), and in the Gulf States to execute a forward strategy of defense.[13] Al-Qaeda and the Islamic State and their legions of foreign fighters have aptly demonstrated the efficacy of nonstate actors in operating in these sorts of operational environments.

LIMITED OFFENSIVE INDIRECT INTERVENTIONS

Three limited UW missions are discussed next: Angola (1975–1990), Nicaragua (1981–1991), and northern Iraq (1991–2002). Why focus on these three cases? Angola and Nicaragua were chosen for their Cold War context. This seems especially salient for today's era of GPC. Iraq during that time period was chosen because it occurred in the post–Cold War period after the 1990–1991 Persian Gulf War.

Angola (1975–1990)

Located in southern Africa, Angola, which is almost twice the size of Texas, is bordered by the nations of the Democratic Republic of the Congo (formerly Zaire), Republic of the Congo, Namibia, and Zambia. It is a nation of over 32 million people today and contains deposits of petroleum, diamonds, iron ore, phosphates, copper, feldspar, gold, bauxite, uranium. It achieved independence from Portugal in November of 1975 and immediately a civil war broke out between the communist People's Movement for the Liberation of Angola (MPLA) and the anti-communist National Union for the Total Independence of Angola (UNITA).[14] The first phase of this civil war, which we are interested in here, lasted from 1975 to 1991. The war became a proxy contest between the United States and South Africa supporting UNITA and the Soviets and Cubans supporting the MPLA.

Even during the Détente era, the United States was concerned about regional balances of power and the spread of communist regimes. Pre-independence, in late January 1975, the CIA proposed to support Alvaro Roberto, the leader of the National Liberation Front of Angola (FNLA), with $300,000 in political action funds. President Ford and his high-level intelligence review panel, known as the Forty Committee, approved this proposal.[15] By July of that year the Forty Committee approved a $14 million two-step program of arms and other aid to support Roberto and Jonas Savimbi, the leader of UNITA.[16] Later that month, President Gerald Ford signed off on the program.[17] A month later, the CIA sent in weapons and eighty-three paramilitary and logistics officers to assist the FNLA and UNITA in Angola in their fight against the MPLA.[18]

Figure 4.1. Map of Angola. *Source*: University of Texas, "Angola," University of Texas, https://legacy.lib.utexas.edu/maps/africa/angola.gif.

(Throughout the 1970s, the Soviet Union was attempting, and in many cases succeeding, in expanding communist regimes across what was then referred as the Third World. South Africa was seen at the time as a stalwart pro-Western regime and it faced a security dilemma across the Frontline States—Angola, Botswana, Mozambique, Tanzania, Zambia, and eventually Zimbabwe—of southern Africa. Supporting South Africa in Angola was seen as part of the larger Cold War geopolitical contest.[19])

In the fall, things would start to ramp up. In September of that year, Congress approved the provision of $10.7 million in aid approved to Angolan rebel forces.[20] The next month, South Africa, supported by the CIA, launched Operation Savannah in support of UNITA and the FNLA against the military wing of the MPLA, who were in turn supported by Cuban forces and Russian advisors. By mid-fall of that year, the U.S. covert action accounted for nearly $23 million in the provision of covert aid, propaganda, weapons, and communications gear.[21]

In November the CIA reported on the situation in Angola and proposed options toward deterring Soviet influence in Angola as well as toward fostering a peaceful environment. These suggestions included detente with the Soviet Union, an arms embargo against Angola, and recognition of an FNLA-UNITA government. Secretary of State Henry Kissinger then allegedly instructed the CIA to take the necessary steps to win the war in Angola. This in turn prompted a closer CIA alliance with South Africa.[22] Within the following days, the Forty Committee and President Ford authorized an additional $7 million to Angola, which made covert aid to Angola in 1975 total $32 million.[23]

But things were about to change. On December 5, 1975, Bill Nelson, the CIA deputy director of operations, admitted in front of the Senate Foreign Relations Committee that the CIA deliberately misled Congress as to its doings in Angola.[24] Following these revelations, Senators Dick Clark, John Tunney, and Alan Cranston introduced an amendment to the U.S. Arms Control Act, which prohibited the United States from providing covert defense funds to Angola without the permission of Congress.[25] This Clark Amendment was signed into law in February 1976.[26] Operation Savannah officially ended in March 1976. The amendment and the election of President Jimmy Carter led to a halt in U.S. covert action.

The lack of U.S. assistance and increased Soviet and Cuban aid to the MPLA caused UNITA to pursue more brutal operations against the civilian populace.[27] By the end of 1976, the CIA had estimated that between 13,000 and 18,000 Cuban troops had deployed to Angola.[28] As with the earlier FID cases examined, there can be a negative impact in not monitoring the activities of supported forces, whether they are governmental forces or insurgents.

The election of President Ronald Reagan in 1980 would lead to a policy of rollback against communist regimes. Throughout his presidency, the administration encouraged South African involvement in Angola through direct intervention and provision of supplies. Additionally, Reagan defied the Clark Amendment and provided direct funds to UNITA. Both of these factors played a role in the rebuilding of UNITA.[29] In January 1982, Jonas Savimbi, the leader of UNITA, gave an interview to the Portuguese weekly newspaper *Tempo* in which he publicly implied that the USG is supplying UNITA with material aid in defiance of the Clark Amendment. "The American government intends to create a dynamo in southern Africa, in Namibia, in Angola. . . . If they want to give assistance, they can always give it. The Clark Amendment does not mean anything."[30] Finally, on June 11, 1985, the Republican-controlled Senate repealed the Clark Amendment. Following this repeal, UNITA forces were given $13 million from the USG.[31]

By 1986, U.S. covert aid to UNITA through Zaire was officially acknowledged, initially starting at $15 million and rising to around $50 million by 1991. However, throughout Reagan's presidency (1981–1989), aid never rose above $30 million.[32] On February 10, 1986, Reagan ordered National Security Decision Directive (NSDD)-212, which stated that the United States would remain involved in Angola in order to achieve several objectives, including the elimination of Soviet and Soviet-proxies (Cuba) from Angola and South Africa; the promotion of regional stability; and the guidance of African countries away from apartheid.[33] *Africa Today* in May 1986 reported that the United States had been supplying UNITA with weapons through intermediaries in Saudi Arabia, Israel, and Zaire.[34]

More controversy arrived in July 1987 when Sam Bamieh, a U.S. businessman, reported to a congressional investigating committee that the Saudi government requested him to channel more than $50 million to UNITA in 1983 as part of the Airborne Warning and Control System sale of five of those aircraft to Saudi Arabia. Bamieh stated that William Casey, the director of Central Intelligence, had solicited the aid.[35]

In 1988, the CIA budget for UNITA was estimated at $30–$45 million.[36] It was increased to $50–$60 million in 1989.[37] This increase in aid followed a promise by president-elect George H. W. Bush to Savimbi to aid the Angolan rebels. Of course, by this point it was estimated that there were 50,000 Cuban troops within Angola and the USSR was supplying $1.5 billion per year in aid to the Angolan government.[38] But the shifting geostrategic chessboard and the human and financial costs to Havana had them begin to withdraw forces in January of that year—a process that was completed by summer.

While the geopolitical stakes on the ground diminished, the United States continued to hold the Angolan government's feet to the fire in 1990. In July of that year, the United States voted to reject Angola's application for membership in the World Bank and the International Monetary Fund.[39] Later that year, in September, the State Department urged Savimbi to attend a new summit in Zaire. Despite not attending the summit, Savimbi still garnered U.S. support. Additionally, the State Department "leaned" on Mobutu Sese Seku, then president of Zaire, to maintain its military supply link to UNITA and issued a statement supporting UNITA's negotiating stance.[40] But in November Congress cut off all aid to the Mobutu regime.

The end of the Cold War and the withdrawal of South African troops from neighboring Namibia eventually allowed for a peace deal and elections in

1992. Absent the geostrategic component, the United States pulled its support. But the civil war would continue shortly after the 1992 election until 1994. And then after four years of peace, the war would re-erupt from 1998 to 2002. Over 500,000 Angolans were killed in the fighting.

U.S. covert support of the insurgency in Angola cannot be decontextualized from the Cold War era. For a relatively small investment, the United States was able to cause both the Soviets and Cubans to expend significantly more blood and treasure in the conflict. In strictly realpolitik fashion, this intervention fit within U.S. strategic objectives. Morally, however, the situation on the ground was more chilling.

Nicaragua (1981–1991)

Nicaragua is located in Central America bordered to the north by Honduras and to the south by Costa Rica. The Caribbean Sea borders it to the east and the Pacific Ocean to the west. It is slightly larger than the U.S. state of Pennsylvania with a population of 6.2 million people and its natural resources include gold, silver, copper, tungsten, lead, zinc, timber, and fish. The country declared independence from Spain in 1821 and became an independent country in 1838. During the first half of the nineteenth century, Great Britain occupied the Caribbean coast. In 1978, a civil war erupted over government manipulation and corruption. This civil war led to a civic-military coalition driven by the Marxist Sandinista movement and its leader Daniel Ortega Saavedra to power in 1979.[41]

Much like in the case of Angola, the election of President Reagan caused a shift in U.S. policy. In April 1981, his administration terminated all economic assistance to the Nicaraguan government due to its involvement in supporting Salvadoran communist insurgents. Reagan attempted to dissuade Managua from supporting the insurgents but eventually decided to fund a clandestine guerilla force (mainly composed of ex-national guardsmen who had gathered in Honduran territory) known as the Contras. The Contras mission revolved around stopping the Sandinistas from training Salvadoran rebels.[42]

On December 1, 1981, President Reagan signed off NSDD-17, which authorized the CIA to recruit and aid the Contras and to block Cuba and Nicaragua's support of insurgences in Central America.[43] This led to $19 million being provided to the Contras in support of stopping the flow of weapons from Nicaragua to El Salvador and bringing down the Sandinista government. Operations included intelligence gathering and paramilitary actions within Nicaragua. The CIA received pledges from Honduras and Argentina to build bases and train rebels.[44]

As in Angola, the Reagan administration selected Nicaragua as a target for rolling back communism.[45] In 1983, Congress allocated another $19 million in covert aid to the Contras, but prohibited it from being used as a means to

Figure 4.2. Map of Nicaragua. *Source*: University of Texas, "Nicaragua," https://legacy.lib.utexas.edu/maps/americas/nicaragua.gif.

topple the Government of Nicaragua. Aid and training continued into 1983. During the second half of 1983, the CIA conducted airstrikes with the Contras on an airport near the capital city Managua.[46] In November of that year, Congress committed another $24 million in assistance to the Contras with the stipulation that only the CIA may utilize the money, not the rebels themselves.

But then Congress became wary. On December 12, 1983, the first Boland Amendment became law. This amendment prohibited both the CIA and DOD from providing funds for the purpose of overthrowing the Nicaraguan government or facilitating military exchange between Nicaragua and Honduras. But it had two significant loopholes that allowed the administration to continue supporting the Contras. First, the CIA had permission to send aid to the Contras as long as they stated its purpose was for anything other than overthrowing the Nicaraguan government. second, the Reagan administration determined that it could use the National Security Council (NSC) to conduct covert operations in Nicaragua rather than the CIA and still remain within the law.[47]

Activity on the ground and at sea ramped up. On January and February of 1984, the CIA placed sea mines in Nicaraguan harbors.[48] Initially the Contras claimed responsibility for placing the mines. That year the administration requested $21 million to finance the Contras, but Congress rejected that request when it learned the CIA emplaced the sea mines and cut off all aid.[49] Seeking ways around congressional restrictions, an NSC staffer, Marine lieutenant colonel Oliver North, solidified an agreement with retired air force Major General Richard Secord and Iranian businessman Albert Hakim to supply the Contras with $11 million in arms via private funds.[50]

During the State of Union Address on February 6, 1985, President Reagan stated, "We must not break faith with those who are risking their lives—on every continent from Afghanistan to Nicaragua—to defy Soviet-supported aggression and secure rights which have been ours from birth."[51] This would later be declared as the so-called Reagan Doctrine. Later that year in June, Congress appropriated $27 million in assistance for items such as food, clothing, medical supplies, and transportation but no military aid.[52] But the next month Lieutenant Colonel Oliver North diverted funds from Iranian arms sales to the Contras.[53] Later known as the Iran-Contra scandal, the idea was to overcharge the Iranian government for weapons and munitions it needed for its war with neighboring Iraq and use the excess funds to fund the rebels in Nicaragua. And throughout that year, North and Secord worked together to build an airstrip in Costa Rica that would allow further supply of arms to the Contras.[54] The airstrip became fully functioning by May of 1986.

Congress raced to close the loopholes. On December 8, 1985, Congress passed the second Boland Amendment. This amendment prohibited any U.S. agencies from providing aid to military or paramilitary forces in Nicaragua and banned the Reagan administration from soliciting funds for the Contras from third-party or private donors.[55]

Despite this, between 1984 and 1986, the NSC raised over $34 million in aid for the Contras through third countries such as Saudi Arabia and conservative fundraisers.[56] North was also able to obtain $1 million in funds from Taiwan. In both 1985 and 1986, Carl R. Chanell, a conservative fundraiser, raised over $12 million from organized fundraisers and allocated $2.7 million to the Contras.[57]

The next year, on February 25, 1986, the president requested $100 million in DOD funds for military and other aid to the Contras, reasoning that the money would be used to stop the growing communist threat in Nicaragua. About 70 percent of aid was designated for weapons (arms, ammunition, Stinger anti-aircraft missiles, training by SF at rebel bases in Honduras) and the remaining for other supplies (food, uniforms, boots, and medical supplies), all to be given to the Contras over an eighteen-month period beginning on March 31, 1986.[58] Congress approved those funds during the summer.

But things became troubling that fall. On October 5, 1986, a plane transporting arms to the Contras was shot down by the Sandinistas, resulting in the capture and imprisonment of Eugene Hasenfus, a U.S. Marine.[59] His statements to the Sandinista government marked the beginning of the exposure of the aforementioned Iran-Contra affair. Still, later that month President Reagan signed into law the bill approving $100 million of military and humanitarian aid to the Contras.[60] Iran-Contra would create a political firestorm.

Still, funding for the Contras continued. By the end of September 1987, the assistance expired but the Contras continued to receive money and materials. Soon afterward, Congress authorized another $20 million in nonlethal aid through stopgap spending bills.

U.S. aims in Nicaragua were to stop the spread of communist revolution in places such as El Salvador, to lead to a change to a representative government in Managua, and to bring peace to the region. The policies led to that, but at a cost. Ortega lost the 1990 presidential election to Violeta Barrios de Chamorro, who was U.S. supported, and her fourteen-party National Opposition Union.

According to Linnington, the U.S. UW campaign was a success, as it

> contributed to achieving U.S. objectives to contain the Sandinista revolution and to bringing about the creation of a representative government. However, it did so with great costs to the Nicaraguan people and to the American presidency. For the Nicaraguan people, the decade of war resulted in the loss of 30,000 lives, or almost 1 percent of the total population. Domestic economic mismanagement coupled with U.S. economic warfare rendered Nicaragua the poorest country in the Western Hemisphere by 1990.[61]

Like the case of Angola, the process was not a pretty one. But U.S. policy cannot be decontextualized from its Cold War context. In the short term. the United States did achieve its objectives. Central America did not become another Vietnam costing a massive intervention by U.S. forces. U.S. conventional capabilities focused on the Soviet and Warsaw Pact threat in Europe and elsewhere while both sides of the Cold War engaged in proxy wars on the periphery.

Northern Iraq (1991–2002)

In August of 1990, Saddam Hussein's Iraq invaded and occupied Kuwait. In response, the global community, led by the United States, organized a coalition of thirty-five nations to deter Iraq from further military action against nations on the Arabian Peninsula and later to expel Iraqi forces and restore Kuwaiti sovereignty. These operations occurred in two phases: the defensive Operation Desert Storm (August 2, 1990–January 17, 1991) and the offensive

Operation Desert Storm (January 17–February 28, 1991). During the second phase, Iraqi forces were routed and driven from Kuwait during the course of these operations but many elite Iraqi forces survived. While it was hoped that Hussein's Baath Party regime would not survive military defeat, this was not the stated aim of the operations and the regime did in fact remain in power.

In 1991, Iraq had a population of approximately 19.5 million people. Ethnically the population was split between the Arab majority (75–80 percent); Kurds (15–20 percent); and Turkomans, Assyrians, or other (5 percent). The vast majority of the population was, and remains, Muslim (97 percent) with a small minority of Christians and other groups (3 percent). But the Muslim population was split between the Shia (60–65 percent) and Sunnis (32–37 percent) religious communities.[62] The fact that the controlling Baath party was predominantly Sunni Arab, coupled with the stinging defeat of the war, led to uprisings of both the southern Arab Shia population and the northern Kurdish minority populations. We will discuss the northern uprising and aftermath here.

The predominantly Muslim Kurdish population had long sought autonomy or independence from Baghdad. A nation without a state, Kurds are spread

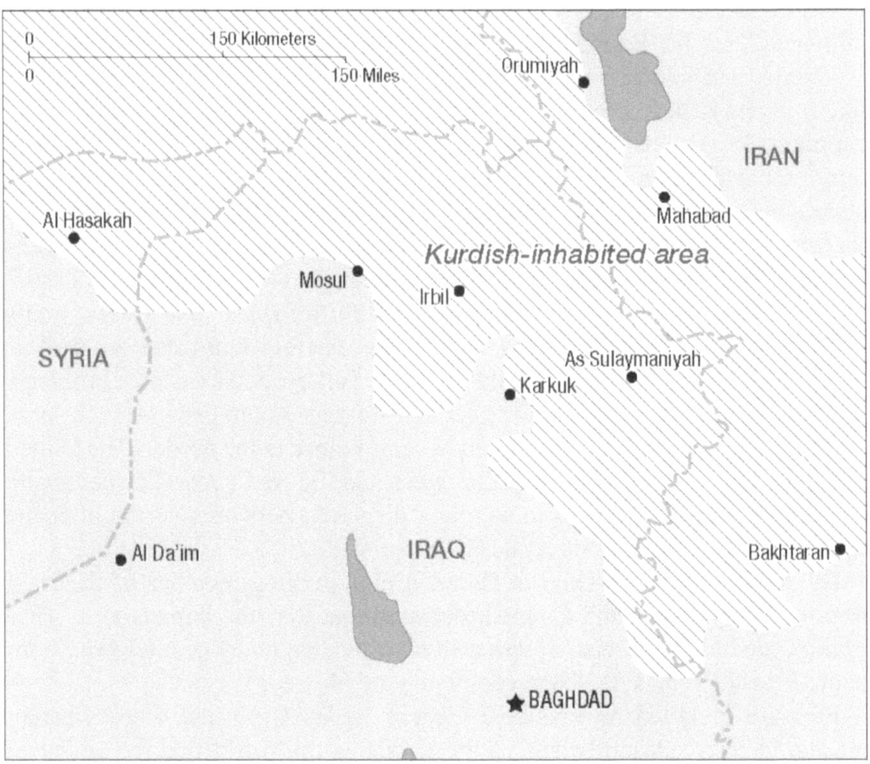

Figure 4.3. Map of Kurdish Areas in Iraq and Neighboring Countries. *Source*: https://legacy.lib.utexas.edu/maps/middle_east_and_asia/kurdish_lands_92.jpg.

out across the countries of Iraq, Iran, Turkey, and Syria—see figure 4.3. Following the aftermath of the 1990–1991 Gulf War, Saddam Hussein ruthlessly went about reestablishing control in both the southern Shia areas and over the Kurdish northern part of the country.

On March 5, 1991, instigated by an uprising in the city of Raniya, Peshmerga (Kurdish militia forces), civilians, and Iraqi military deserters overthrew government control in cities across the Kurdish region. They took control of approximately 75 percent of the region over the course of the next week.[63] Kurdish rebels, encouraged to revolt by CIA radio broadcasts during Operation Desert Storm, likely expected the United States to support the uprising.[64]

A little over a week later, the White House shied away from taking a position on the Iraqi uprisings, arguing that they are an internal matter.[65] But Kurdish operations continued. On March 14, Kurdish forces converged on Mosul and subdued the garrison there.[66] Days later, a Kurdish force of 100,000 men assaulted and captured the oil-rich city of Kirkuk within three days.[67] But then the Iraqi government fought back. From March 20 to 22, Iraqi helicopter gunships and fixed-wing aircraft targeted civilians as well as insurgents in Kurdish cities such as Kirkuk. U.S. Air Force (USAF) fighter aircraft out of Incirlik, Turkey, downed several Su-22 fighter bombers on the 20th and 22nd, but Baghdad's helicopters continued to operate unmolested.[68]

Lacking sufficient firepower and at a geographical disadvantage in cities like Irbil and Kirkuk, the Peshmerga were unable to hold the territory they had captured. Desertions and a lack of international aid compounded the problem.[69] On March 28, in the absence of U.S. support for the rebels, the Iraqi central government launched a sustained offensive to recapture lost towns in the north.[70] By March 30, the Peshmerga were unable to repel Iraqi forces and civilians were forced to flee into the mountains toward Turkey and Iran.[71]

By the beginning of April almost 1 million refugees fled toward Iraq's northern border.[72] Iran accepted some refugees from Kurdistan, as well as many Shiites fleeing from southern Iraq. But Turkey, due to its own problems with its Kurdish population, was reticent to accept refugees in such large numbers and attempted to keep them in camps close to the border. The United Nations High Commissioner for Refugees and the Red Cross/Crescent were hard-pressed to provide aid to so many displaced persons so soon after the Gulf War depleted their stockpiles of supplies.[73]

By April 4, Izzat Ibrahim al-Douri, a high-ranking member of the Iraqi Revolutionary Command Council, declared that the "the homeland . . . has been purged of the rebels." It was also reported that many of the towns sympathetic to the rebels had been completely eliminated.[74]

Pressure mounted on President George H. W. Bush and other Western leaders to intercede on behalf of the Kurds as the world became aware of their plight. Turkey estimated that 220,000 refugees fled toward its borders. Kurds who two months earlier had praised President Bush now accused him of colluding with Saddam in the perpetration of genocide. On April 5, citing letters from Iran, Turkey, and France urging action in the face of humanitarian crisis, the United Nations Security Council (UNSC) adopted Resolution 688 condemning Iraq's actions and demanding immediate access for humanitarian workers.[75] President Bush ordered U.S. forces to eastern Turkey to aid in relief efforts. NATO called for international pressure to be applied against Iraq to stop the repression of Kurds and other civilians.[76]

The next day, Joint Task Force Provide Comfort was formed in Incirlik, Turkey, under the command of Major General James L. Jones and was given the task of supplying humanitarian relief to Kurdish refugees on the border.[77] The United States declared the first no-fly zone in northern Iraq, banning all air activity by the Iraqi military north of the 36th Parallel. USAF and U.S. Navy (USN) aircraft enforced the ban and provided reconnaissance tracking the movements of refugees and Iraqi forces.[78] U.S. forces flying from Germany and Turkey began airdropping supplies to Kurdish refugees in the Iraq-Turkey border region.[79]

On April 13, U.S. soldiers were deployed to the mountains to distribute aid and care for the sick and injured refugees; however, the size of the problem is larger than a limited contingent can handle. Dehydration and diarrhea were rampant throughout the refugee camps.[80] By April 16, the immediate needs of refugees were being met by coalition forces; but they feared returning to their homes. Coalition forces then agreed to establish a security zone in northern Iraq in order to convince the refugees to leave the mountains.[81]

President Bush ordered U.S. troops into northern Iraq to begin helping Kurds out of the mountains and into more manageable terrain in northern Iraq, where it would be easier to provide them with supplies. Because this necessitated the presence of U.S. personnel in Iraq, the operation developed a security requirement in addition to its core mission. In a press conference announcing the decision, the president reaffirmed the objective of Operation Provide Comfort as assuring that "all Iraqi refugees are able to return home and live in peace."[82]

Now commanded by Army LTG John Shalikashvili (later GEN and chairman of the Joint Chiefs of Staff), Operation Provide Comfort spent the rest of the year establishing security to allowing the Kurdish population to return home. By September, Kurdish guerillas enjoyed large success in fighting off any encroachments by Baghdad's forces due in part to the level-playing field provided by the no-fly zone. But the shelling of several cities forced 200,000 Kurds to flee toward Iran.[83] And by October, the weakened state of Iraqi forces after the Gulf War defeat and the attention being paid to the situation prevented Baghdad from engaging the Peshmerga directly. Instead, Iraqi forces blockaded Iraqi Kurdistan in an attempt to cripple the economy there.[84]

Aid operations continued throughout the next year as well as enforcement of the no-fly zone. On May 19, 1992, elections were held in Kurdistan resulting in a virtual draw between the two major parties the Kurdish Democratic Party (KDP) and the Patriotic Union of Kurdistan (PUK).[85] Together they agreed to seek autonomy within Iraq rather than try to achieve independence—this reassured Turkey and other neighboring states who feared the precedent such agitation would set for their own, native Kurdish populations.[86]

The Kurdistan National Assembly attempted to create a unified command of 80,000 Peshmerga to replace the existing faction-controlled units; however, neither the KDP nor the PUK were willing to relinquish control of their forces. A unified command was established in September, but ultimate control remained fractured.[87] In November, William Jefferson Clinton was elected as president of the United States.

Operation Provide Comfort continued throughout 1992 and 1993—it would go on until December 1996 when it was replaced by Operation Northern Watch (1997–2003). Various uses of force occurred, including attacks on Iraqi aircraft, air defense sites, and airfields. On June 26, 1993, President Clinton ordered a strike of twenty-three cruise missiles on the Iraqi intelligence headquarters outside of Baghdad in response to that agency's April assassination attempt on former president Bush during his visit to Kuwait.[88]

But by early 1994, the domestic political situation in Iraqi Kurdistan had become tenuous. The PUK-KDP coalition government collapsed into armed clashes over territorial control and revenue sharing.[89] KDP forces outnumbered their PUK counterparts; however, the PUK possessed superior heavy equipment, including tanks and artillery.[90]

U.S. military forces and the CIA continued to build relationships with Kurds on the ground. The de facto Kurdish autonomous region supported by the no-fly zone April allowed for continued coordination and relationship-building. But on April 14, tragedy struck when twenty-six people from four coalition militaries and the Kurdish government were killed when two U.S. Army Blackhawk helicopters were shot down by USAF F-15s in a friendly fire incident over the no-fly zone.[91] The F-15s visually misidentified the Blackhawks as Russian-made Iraqi helicopters after being unable to determine their identity electronically and fired on them.[92]

The civil war between the PUK and KDP continued into 1996. In August of that year, in an attempt to gain the upper hand over the PUK, the KDP invited Saddam Hussein to retake Irbil. Iraqi forces seized the lightly defended city and drove the PUK forces back toward the Iranian border.[93] President Clinton responded by expanding the southern no-fly zone to the 33rd Parallel and striking more air defense targets within Iraq.

In response to the retaking of Irbil, the last U.S. military ground presence in Iraq—a Special Forces detachment stationed in Zakho since 1991—left the country.[94] The Iraqi government-KDP alliance also caused the United States to withdraw CIA/DOD northern Iraq liaison element teams from the region. (These teams had worked with Kurdish forces to gather intelligence and undermine the regime in Baghdad politically. At one point, the CIA backed a plot to overthrow Saddam Hussein with Kurdish and rogue Republican Guard forces. The inter-Kurdish rivalry had long frustrated the CIA.[95]) And on September 3, 1996, the United States initiated Operation Desert Strike with the launch of cruise missile attacks against Iraq in retaliation for the siege of Irbil.[96] In addition, the United States initiated Operation Quick Transit, which evacuated approximately 6,000 endangered Kurds who had supported American operations from the region.[97]

By September 1998, U.S.-backed negotiations between the KDP and PUK led the two sides to issue the Washington Declaration, which ended four years of conflict between the two parties. (The agreement was formalized when the Kurdish parliament reconvened in 2002.[98]) And on October 31, President Clinton signed the Iraq Liberation Act of 1998, which affirmed the U.S. position that Saddam Hussein should be removed from power in Iraq and pledged support for democratic movements within the country. The act would later be cited as justification for the 2003 invasion of Iraq.

Throughout the 1990s, both the CIA's Iraq Operations Group and American SOF worked to overthrow the Saddam Hussein regime and developed the networks to conduct a UW campaign in northern Iraq, respectively.[99] But these efforts were episodic at best and tragicomic at worst.[100] Resources were shrinking, political will in the U.S. leadership for robust regime change was lacking, and tensions, bloodletting, and double-crossing between the KDP and PUK factions conspired against success. U.S. efforts were more aligned with keeping the Baghdad regime in check and not allowing it to threaten or invade its neighbors. On that score, the operations within Iraq, when coupled with regional shows of force and strikes such as 1998's Operation Desert Fox, worked. But 9/11 and concerns over the possibility of Saddam Hussein's regime possessing or using chemical or biological weapons would lead to a more robust response and regime change in 2003.

ASSERTIVE OFFENSIVE INDIRECT INTERVENTIONS

While it is theoretically possible that the United States could carry out assertive offensive indirect interventions independently from a broader operation, the recent historical record shows no such cases. This section therefore covers the use of CIA elements and SOF as part of the larger campaigns in the invasions of Afghanistan (2001) and the campaign in northern Iraq (2002–2003). Both of these interventions illustrate the capabilities that these forces bring to the table and their ability to achieve success far greater than their numbers might suggest. Of course, in both instances these forces did not work alone, they were effective because they were working with and through local forces on the ground. Furthermore, both cases show that as brilliant as initial tactical and operational effectiveness may be, that does not necessarily lead to lasting strategic gains.

Afghanistan (2001)

Afghanistan, a poor mountainous country in Central Asia, has had a long and messy history. In the contemporary period, the country has had few years of peace. Civil strife in the 1970s led to a Soviet invasion in 1979. That external intervention in turn led to a response from nations such as neighboring Pakistan, from co-religionists such as Saudi Arabia, and from the USSR's strategic rival, the United States. It also brought in an influx of foreign volunteers from across the Muslim world. Among these foreign fighters was the scion of a Saudi construction magnate named Osama bin Laden.

While the Soviet-Afghan war led to indirect interventions by many sides, eventually the cost in blood and treasure led the Soviets to withdraw. Rather than leading to peace, however, this withdrawal caused more death and destruction as the country fell into civil war. Bloody fighting from 1989 to 1996 led to the establishment of an austere Islamic government under the banners of the Taliban movement and their leader, Mullah Omar.

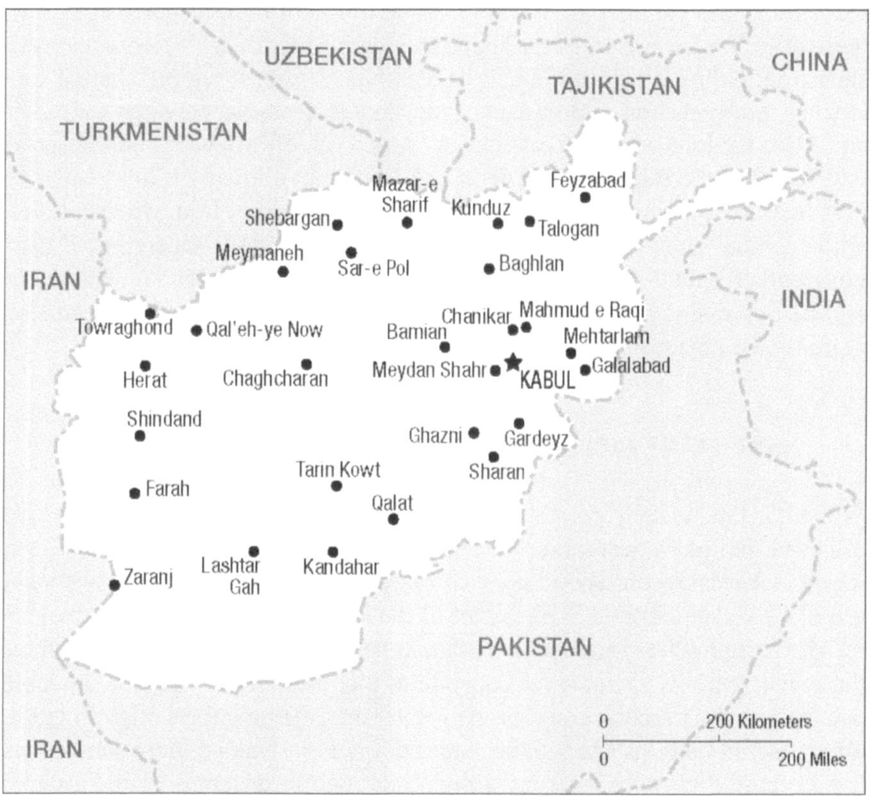

Figure 4.4. Map of Afghanistan. *Source*: http://legacy.lib.utexas.edu/maps/middle_east_and_asia/afghanistan_pol01.jpg.

The establishment of an Islamist government was long viewed as an objective of various factions who had battled the Soviet occupation in Afghanistan. A confluence of circumstances led a now older Osama bin Laden to return to his old battleground after he had worn out his welcome back home in Saudi Arabia and then, under U.S. and Saudi pressure, from his exile home in Khartoum, Sudan. His distaste for both the Saudi royal family and U.S. policy had led him the create the secretive al-Qaeda movement. His new base of operations allowed him the space and ability to train groups and individuals for operations against what he saw as apostate Muslim regimes (the "near enemy") and what he saw as their backer, the United States and the West (the "far enemy").

The cadre and followers of his movement, and other like-minded groups and individuals, came to Afghanistan. From this out-of-the-way location, plots and plans were conceived and individuals were trained to carry out operations. Fearlessly he declared war on his enemies and coordinated attacks on such targets as the East African U.S. embassies in Nairobi, Kenya, and Dar es Salaam, Tanzania (1998), and on the USS *Cole* in Yemen's Aden harbor (2000).[101] The embassy bombings led to cruise missile strikes in 1998 and brought the attention of the CIA's SAD and U.S. SOF, but no successful operations were conducted over political objections in the United States and on a lack of consistent actionable intelligence.[102]

Bin Laden miscalculated in thinking that the United States was a paper tiger. On September 11, 2001, nineteen hijackers using boxcutters hijacked four domestic passenger planes. Aircraft struck each of the Twin Towers in lower Manhattan and the Pentagon in Washington, DC. The fourth plane was brought down in a field in western Pennsylvania when the passengers fought back. The attacks killed 2,996 people (including the hijackers), wounded more than 6,000, and caused billions of dollars in damage. The United States was at war. Hedging his bets in case the United States did respond, bin Laden had the most capable Afghan leader of the opposition United Front, Ahmad Shah Massoud, assassinated in a suicide bombing attack on September 9, 2001.

President George W. Bush called upon the Taliban to extradite Osama bin Laden and those responsible for the attacks. The Taliban refused. On September 18, 2001, the U.S. Congress in a joint resolution authorized the use of military force. The resolution stated:

> That the President is authorized to use all necessary and appropriate force against those nations, organizations, or persons he determines planned, authorized, committed, or aided the terrorist attacks that occurred on September 11, 2001, or harbored such organizations or persons, in order to prevent any future acts of international terrorism against the United States by such nations, organizations or persons.[103]

Now the United States just had to figure out how it would do so. For the first time in the history of NATO Article 5 was invoked for collective defense—that is, an attack on the United States constituted an attack on all of the allies.[104]

While the Pentagon struggled with a plan, the CIA were the first to offer something satisfactory to the president. Agency SAD personnel, under the direction of Gary Schroen, were the first boots on the ground on September 26.[105] Carrying $3 million, Schroen's team linked up with opposition forces to buy support for U.S. efforts, pave the way for follow-on Special Forces, and enable real-time communication about the situation on the ground. Aerial bombing of targets in Afghanistan commenced on October 7.

U.S. military efforts for Operation Enduring Freedom-Afghanistan (OEP-A) were led by GEN Tommy Franks, the CINC of U.S. Central Command, which had responsibility for Afghanistan in its AOR. (CINCs would later be known as Regional Combatant Commanders because Secretary of Defense Donald Rumsfeld felt that there should be only one CINC, the president.) U.S. forces would fall under various subcomponent commands, which included a mix of SOF and conventional forces. The United States, however, was not alone. Britain sent members of the SAS and Royal Marine Commandoes Special Boat Service and intelligence officers from MI-6 to fight alongside the Americans. The Australians also deployed members of their SAS Regiment, and the Canadians deployed members of their elite Joint Task Force-2.

The initial 5th SF Group ODAs (twelve personnel cross-trained in various military skills) with U.S. Air Force Combat Controllers (CCTs; a type of unit within the USAF's SOF community) were inserted to link up with Northern Alliance opposition forces on October 18.[106] Follow-on ODAs with CCT personnel linked up opposition forces across the country. The CCTs were able to call in the complete inventory of plane-delivered ordnance that helped to wear down the capabilities and numbers of Taliban and al-Qaeda Brigade 055 forces. Direct action forces of the United States and allied SMUs also began to conduct operations.

Slowly Northern Alliance forces backed by United States and allied support began to take ground. By November it seemed possible that Kabul could be taken. But cracks began to develop between coalition and our Afghan allies on the ground. The Northern Alliance wanted to finish off the Taliban, while the United States wanted to continue to focus on al-Qaeda. On November 13, Kabul fell. Around the same time Taliban and al-Qaeda forces began to concentrate on an old mujahideen base named Tora Bora that was close to the city of Jalalabad. U.S. airstrikes, SOF, and SAD personnel started to attack the area on November 16. It is possible that Osama bin Laden was there. If he was he slipped away, and would not find justice until the U.S. SMU raid on Abbottabad, Pakistan, on May 2, 2011.

Operations continued on the ground and in the air, but the conflict would soon take more of a conventional tilt. The war in Afghanistan would soon turn into a long slog. The coming war in Iraq would divert forces and resources from the conflict and the local situation on the ground with various elements of the Taliban, al-Qaeda, and flocking foreign fighters from Pakistan and the rest of the Islamic world becoming insurgents and a weak central government in Kabul under the leadership of soon-to-be-elected President Hamid Karzai would prove to be an unreliable partner. At the time of this writing U.S. forces remain in Afghanistan while peace talks are continuing with the Taliban and the United States proposing a pullout of most of its forces.

Could the war in Afghanistan have taken a different turn? It is certainly possible, but that is a big what if? Had Osama bin Laden and other senior leaders of al-Qaeda been killed or captured then perhaps that would have been enough to leave quickly and not get involved in the internal Afghan contests. Retired Army colonel Hy Rothstein has argued that the United States turning the campaign into a more conventional campaign with a more centralized command and control structure hindered the ability of SOF and others to achieve more progress.[107] While that is possible, it also discounts the divergence between U.S. and Northern Alliance interests during the early months of the war. As was discussed earlier in this chapter, that is a classic problem in proxy war environs. The United States did not have the forces needed to impose its will and the introduction of such forces would have posed serious logistical issues and likely would not have been welcomed by the local populace. There were also issues with squabbles between SOF and the CIA over information sharing and the ability—and authority—to distribute funds to indigenous forces on the ground.[108] (The latter issue would later be eased by what is now known as 127e authority, which was discussed in a previous chapter.) In the end, the promising start to OEF-A would soon lose its luster.

Northern Iraq (2002–2003)

The George W. Bush administration took office with a tougher stance toward Iraq than the Clinton administration had taken. After 9/11, this stance was hardened further. The administration tried to ascertain whether there were links between Saddam Hussein's regime and al-Qaeda. Concerns were also raised about Iraq trying to escape the international sanctions regime that was in place and on Saddam Hussein's possible possession of weapons of mass destruction (WMD). The president decided in the summer of 2002 to go to the UNSC to pressure Iraq to allow in weapons inspectors. In September UNSC Resolution 1441 was passed to authorize weapons inspectors to return to Iraq, with serious consequences if they were not allowed. Following this the U.S. Congress passed an Authorization for the Use of Military Force against Iraq that allowed the president to use force to "defend the national security of the United States against the continuing threat posed by Iraq" and to "enforce all relevant United Nations Security Council resolutions regarding Iraq."[109] On November 13 of that year, Saddam Hussein accepted the resolution and inspectors began to return.

Prior to that, however, the United States had infiltrated SAD and Army 10th Special Forces Group (Airborne) (10SFG(A)) Advance Force Operations personnel in July 2002.[110] The purpose of this insertion was to link up with the Peshmerga, build intelligence networks, and prepare the way for additional SF should an invasion be necessary.[111] SAD personnel also conducted operations in non-Kurdish areas to identify leadership targets and to convince Iraqi commanders to not fight the United States in case of invasion.[112] By December of 2002, the defense publishing outlet *Jane's* reported that the Bush administration authorized the disbursement of $92 in military assistance for equipment, education, and training to the following groups: Iraqi National Accord, Iraqi National Congress, Kurdish Democratic Party, Movement for Constitutional Monarchy, Patriotic Union of Kurdistan, and Supreme Council of the Islamic Revolution in Iraq.[113]

By February of 2003, the die had been cast. Secretary of State Colin Powell addressed the UNSC on February 5. For some his presentation was an ironclad case for war based on information that an Islamist terrorist group, Ansar al-Islam, which comprised local Kurdish recruits and Arab veterans of Afghanistan (including the infamous Abu Musab al-Zarqawi), was producing chemical weapons in Iraq and that the Iraqi government had been violating United Nations Security Council Resolutions (UNSCRs).[114] The United States would go to war the next month.

A fly in the ointment of U.S. war plan was that NATO ally Turkey would not allow the further staging of U.S. forces from its territory. In the original plan the mechanized 4th Infantry Division was supposed to push down from Turkey. U.S. and U.K. SOF and SAD, which had been on the ground for several months by this time, worked with the Peshmerga to tie down as many Iraqi divisions as they could in the north so that they could not be sent south to reinforce against the main thrust of U.S. and British forces coming up from Kuwait. When hostilities broke out, these forces began directing U.S. airstrikes on Iraq's front-line positions outside Mosul and Kirkuk. Their efforts allowed the Kurdish Peshmerga to advance closer to these two cities.[115] On March 26, these forces were reinforced when the Italy-based, U.S. Army's 173rd Airborne Brigade parachuted in near Kirkuk and were later joined by units from the 10th Mountain Division's 2–14 Infantry battalion.[116]

From March 28 to 30, 2003, a force comprising SAD, 10SFG(A), and Peshmerga attacked the Ansar al-Islam camp and suspected Ricin and chemical weapons facility in Sargat, Iraqi Kurdistan (close to the Iranian border), in Operation Viking Hammer.[117] The fighting was quite intense and included a preparatory barrage of over sixty cruise missiles and attacks by AC-130 Spectre gunships. The camp was destroyed but some, like Abu Musab al-Zarqawi the eventual leader of AQI and the forefather of the Islamic State, got away.

U.S. forces eventually toppled the Saddam Hussein regime. But as in the case with Afghanistan, the initial victory was fleeting. The Iraqi insurgency of disaffected Baathists, Shia Arabs, and inflowing foreign fighters would take the liberated Iraq into a prolonged civil war. Large numbers of U.S. troops, and some British forces, remained in the country until 2011.

Were UW operations in northern Iraq a success? In terms of military effectiveness, yes it has to be considered a success. A relatively small number of CIA and SF initially worked with Kurdish Peshmerga forces in order to tie down Iraqi Army divisions in the north. The northern campaign in Iraq was a textbook illustration of the principle of economy of force. The introduction of the 173rd Airborne Brigade added to this task. The previous decade-plus of CIA and SF preparation of the eventual battlefield surely helped. Relationships had been made and useful information gathered that aided the task and contributed to the short-term victory.

CONCLUSIONS

Through the illustration of the five cases presented in this chapter it can be concluded that the United States has achieved some modicum of success in implementing UW campaigns of both the limited and assertive varieties. With that having been said, one cannot overstate the magnitude and duration of such successes. While UW tools can be effective in achieving U.S. short- to medium-term objectives, they are not a silver bullet and are often a messy tool—particularly for the local populaces.

The limited offensive indirect interventions of the Cold War period examined here (Angola and Nicaragua) and the post–Cold War era (northern Iraq, 1991–2002) showed that a small investment of personnel and resources compared to the deployment of a massive U.S. force could achieve government objectives, but they also showed the tensions of working through and by proxies. The assertive indirect intervention cases of Afghanistan (2001) and northern Iraq (2002–2003) also showed that a limited number of intelligence and SOF personnel when combined with precision-guided munitions could offer a more resource-effective way of conducting offensive operations. But they also show that tactical and operational successes do not guarantee strategic-level victory.

Furthermore, as analysts like Eliot Cohen and others have pointed out, such initial success might cloud the minds of decision makers and make them overconfident in such capabilities.[118] The past two decades' fascination with the abilities of SOF, particularly with the scalpel-like "Superman" capabilities, shows that this danger is too real. But perhaps on the flipside, the messiness and longer time durations of the "Daniel Boone" are perhaps beneficial because otherwise politicians might become too fond with them. As was the case with the FID tool then, the UW tool is best when not overprescribed.

NOTES

1. *DOD Dictionary of Military and Associated Terms* (Washington, DC: Department of Defense, January 2020), 227. Available at: https://www.jcs.mil/Portals/36/Documents/Doctrine/pubs/dictionary.pdf.

2. For an excellent discussion of undergrounds, see Paul J. Tompkins, Jr., *Human Factors Considerations of Undergrounds in Insurgencies*, Second Edition (Ft. Bragg, NC: U.S. Army Special Operations Command, Assessing Revolutionary and Insurgent Strategies, 25 January 2013). Available at: https://www.soc.mil/ARIS/books/pdf/HumanFactorsS.pdf.

3. Abigail T. Linnington, "Unconventional Warfare in U.S. Foreign Policy: U.S. Support of Insurgencies in Afghanistan, Nicaragua, and Iraq from 1979–2001" (PhD, Dissertation, The Fletcher School of Law and Diplomacy, Tufts University, December 2012), 5.

4. *DOD Dictionary of Military and Associated Terms*, 205.

5. *Ibid.*, 53.

6. *Ibid.*, 35.

7. Michael E. DeVine, *Covert Action and Clandestine Activities of the Intelligence Community: Selected Definitions in Brief* (Washington, DC: Congressional Research Service, Updated June 14, 2019). Available at: https://fas.org/sgp/crs/intel/R45175.pdf.

8. *DOD Dictionary of Military and Associated Terms*, 133.

9. Jan K. Gleiman, "The American Counterculture of War: Supporting Foreign Insurgencies and the American Discourse of War," *Special Operations Journal* 1, no. 1 (2015): 25.

10. Michael Noonan, "Mumford and Guns," *War on the Rocks* (July 24, 2013). Available at: https://warontherocks.com/2013/07/mumford-and-guns/.

11. Andrew Mumford, *Proxy Warfare* (New York: Polity, 2013).

12. *Ibid.*

13. *Counter-Unconventional Warfare White Paper* (Ft. Bragg, NC: U.S. Army Special Operations Command, 26 September 2014), 29–37.

14. Central Intelligence Agency, "Angola," *The World Facebook* (April 6, 2020). Available at: https://www.cia.gov/library/publications/the-world-factbook/geos/ao.html.

15. Stephen R. Weissman, "CIA Covert Action in Zaire and Angola: Patterns and Consequences," *Political Science Quarterly* (The Academy of Political Science, 1979), 282.

16. *Ibid.*, 282–283.

17. *Ibid.*, 282–283.

18. Prexy Nesbitt, "US Foreign Policy: Lessons from the Angolan Conflict," *Africa Today* 39, no. 1/2 (1992): 56.

19. See, for example, Peter W. Rodman, *More Precious Than Peace: The Cold War and the Struggle for the Third World* (New York: Charles Scribner's Sons, 1994).

20. Weissman, "CIA Covert Action," 283.

21. *Ibid.*, 283.

22. Nesbitt, "US Foreign Policy," 56.

23. Weissman, "CIA Covert Action," 284.

24. Nesbitt, "US Foreign Policy," 57.

25. *Ibid.*, 57.

26. *Ibid.*, 57.

27. Wanda L. Nesbitt, "Jonas Savimbi and UNITA's Struggle for Independence: An Application of Mao's Theory of Warfare?." Curriculum Requirement: National Defense University (1997), 8. Available at: http://www.dtic.mil/dtic/tr/fulltext/u2/a442217.pdf8.

28. Central Intelligence Agency, *Soviet and Cuban Intervention in the Angolan Civil War* (Memorandum, March 1977). Available at: https://www.cia.gov/library/readingroom/docs/DOC_0000518406.pdf.
29. William Minter, "The US and the War in Angola," *Review of African Political Economy* (New York: Taylor & Francis: 1991), 136.
30. U.S. Department of State, Diplomatic Cable. (Document 20), 3.
31. Joel Bainerman, *Inside the Covert Operations of the CIA and Israel's Mossad* (Paris: SP Books, 1994), 17.
32. Nesbitt, "US Foreign Policy," 60.
33. The White House, NSC, *National Security Decision Directive (No. 212) U.S Policy Towards Angola*, by the NSC (February 10, 1986). Available at: https://fas.org/irp/offdocs/nsdd/23-2730a.gif.
34. Bainerman, *Inside the Covert Operations*, 17.
35. *Ibid.*, 16.
36. Minter, "The US and the War in Angola," 139.
37. *Ibid.*, 139.
38. Veneble L. Hammonds, Jr., "Who's Winning: Has Foreign Aid Proven Effective in Angola?" (Analytical Study, Air War College, May 1989), 23. Available at: http://www.dtic.mil/dtic/tr/fulltext/u2/a217249.pdf.
39. *Ibid.*, 139.
40. *Ibid.*, 139.
41. Central Intelligence Agency, "Nicaragua," *The World Factbook* (March 17, 2020). Available at: https://www.cia.gov/library/publications/resources/the-world-factbook/geos/nu.html.
42. "Milestones: 1981–1988—Office of the Historian." Available at: https://history.state.gov/milestones/1981-1988/central-america (accessed April 15, 2019).
43. The White House, NSC, *National Security Decision Directive (no. 17) on Cuba and Central America* (January 4, 1982). Available at: https://fas.org/irp/offdocs/nsdd/nsdd-17.pdf.
44. History.com Editors, "Reagan Gives CIA Authority to Establish the Contras," *HISTORY*. Available at: https://www.history.com/this-day-in-history/reagan-gives-cia-authority-to-establish-the-contras (accessed April 16, 2019); "Understanding the Iran-Contra Affairs—The Iran-Contra Affairs." Available at: https://www.brown.edu/Research/Understanding_the_Iran_Contra_Affair/iran-contra-affairs.php (accessed April 16, 2019).
45. "Understanding the Iran-Contra Affairs—The Counterrevolutionaries (The Contras)."
46. *Ibid.*
47. Elizabeth R. Purdy, "Boland Amendment | Definition, Edward Boland, & Anti-Communist Contras," *Encyclopedia Britannica*. Available at: https://www.britannica.com/topic/Boland-Amendment (accessed April 15, 2019).
48. "Understanding the Iran-Contra Affairs."
49. Special to *The New York Times*, "U.S. Aid to the Contras: The Record Since '81," *The New York Times* (March 20, 1986), sec. World. Available at: https://www.nytimes.com/1986/03/20/world/us-aid-to-the-contras-the-record-since-81.html.
50. *Ibid.*

51. Chester Pach, "The Reagan Doctrine: Principle, Pragmatism, and Policy," *Presidential Studies Quarterly* 36, no. 1 (March 2006): 75.

52. "Milestones: 1981–1988—Office of the Historian."

53. "The Iran-Contra Affair | American Experience | PBS." Available at: https://www.pbs.org/wgbh/americanexperience/features/reagan-iran/ (accessed April 15, 2019).

54. *Ibid.*

55. Purdy, "Boland Amendment."

56. *Ibid.*

57. "Understanding the Iran-Contra Affairs—The Counterrevolutionaries (The Contras)."

58. Special to *The New York Times*, "U.S. Aid to the Contras."

59. "Understanding the Iran-Contra Affairs."

60. History.com Editors, "U.S. Aid to Contras Signed into Law," *HISTORY*. Available at: https://www.history.com/this-day-in-history/u-s-aid-to-contras-signed-into-law (accessed April 15, 2019).

61. Linnington, *Unconventional Warfare in U.S. Foreign Policy*, 159.

62. *CIA World Factbook* (1991). Available at: http://www.gutenberg.org/cache/epub/25/pg25-images.html.

63. Michael G. Lortz, *Willing to Face Death: A History of Kurdish Military Forces—the Peshmerga—from the Ottoman Empire to Present-Day Iraq* (PhD Dissertation, Florida State University, 2005): 59.

64. Bruce A. Weber, *Combined Task Force Provide Comfort: A New Model for 'Lead Nation' Command?* (Newport, RI: U.S. Naval War College, May 16, 1994): 5.

65. Lt. Col. David E. Clary, *Operation Provide Comfort: A Strategic Analysis* (Maxwell AFB, AL: Air War College, April 1994): 4.

66. Lortz, "Willing to Face Death," 59.

67. *Ibid.*, 60.

68. Maj. George D. Kramlinger, USAF, *Sustained Coercive Air Presence: Provide Comfort, Deny Flight, and the Future of Airpower in Peace Enforcement* (Maxwell AFB, AL: Air University Press, February 2001): 20.

69. Lortz, "Willing to Face Death," 60.

70. Clary, Operation Provide Comfort "Strategic Analysis," 4.

71. Kramlinger, "Sustained Coercive Air Presence," 20.

72. Gordon W. Rudd, *Humanitarian Intervention: Assisting the Iraqi Kurds in Operation PROVIDE COMFORT, 1991* (Washington, DC: Department of the Army, 2004): 31.

73. *Ibid.*, 36.

74. Clary, Operation Provide Comfort "Strategic Analysis," 5.

75. UNSCR 688. Accessible at: http://unscr.com/en/resolutions/688.

76. LTC John P. Cavanaugh, *Operation Provide Comfort: A Model For Future NATO Operations* (Ft. Leavenworth, KS: School of Advanced Military Studies, August 25, 1992): 7.

77. *Ibid.*, 8.

78. Kramlinger, "Sustained Coercive Air Presence," 21.

79. Rudd, *Humanitarian Intervention*, 40.

80. *Ibid.*, 21–22.

81. Cavanaugh, "Operation Provide Comfort," 10.
82. Weber, "Combined Task Force," 12–13.
83. *Ibid.*, 26.
84. Lortz, "Willing to Face Death," 62.
85. The KDP was founded by Mulla Mustafa Barzani following World War II and is seen as the first major Kurdish dissident movement. The KDP fought against Baghdad's rule intermittently throughout the twentieth century, sometimes with the support of Iran. In 1964, the group splintered, and gave rise to the offshoot, PUK. The two groups constitute the major political force within Kurdistan and vie against one another for political power and economic resources. See, CRS Report, Jim Zanotti and Clayton Thomas, "The Kurds in Iraq, Turkey, Syria, and Iran," CRS In Focus (Updated January 23, 2019), 1.
86. *Ibid.*, 26.
87. Lortz, "Willing to Face Death," 62.
88. *Ibid.*, 28.
89. CRS Report, *The Kurds*, 2.
90. Lortz, "Willing to Face Death," 63.
91. Capt. Gregory Ball, USAFR, 1991-Operation Provide Comfort and Northern Watch (September 18, 2012). Accessible at: https://www.afhistory.af.mil/FAQs/Fact-Sheets/Article/458953/1991-operation-provide-comfort-and-northern-watch/.
92. Eljay B. Bowron, *OPERATION PROVIDE COMFORT: Review of U.S. Air Force Investigation of Black Hawk Fratricide Incident* (Washington, DC: GAO, June 18, 1998).
93. Daniel L. Haulman, "*Crisis in Iraq: Operation PROVIDE COMFORT,*" *in Short of War: Major USAF Contingency Operations*, 1947–1997 edited by A. Timothy Warnock (Maxwell AFB: Air University Press, 2000): 185.
94. Michael A. Schiesl, *The Objectives of United States Military Intervention in Northern Iraq Between Operation Desert Storm and Operation Iraqi Freedom* (Ft. Leavenworth, KS: Master's Thesis, U.S. Army Command and General Staff College, 2003): 5.
95. Lortz, "Willing to Face Death," 64.
96. Schiesl, "Objectives of United States Military Intervention," 5.
97. *Ibid.*, 5.
98. Zanotti and Thomas, *The Kurds*, 2.
99. See Linnington, *Unconventional Warfare in U.S. Foreign Policy*, chapter 7.
100. See, for instance, Robert Baer, *See No Evil: The True Story of a Ground Soldier in the CIA's War on Terrorism* (New York: Crown Publishing Group, 2002).
101. See, for instance, Steve Coll, *Ghost Wars: The Secret History of the CIA, Afghanistan, and Bin Laden, from the Soviet Invasion to September 10, 2001* (New York: Penguin, 2004); and Lawrence Wright, *The Looming Tower* (New York: Knopf, 2006).
102. See, for instance, Richard Shultz, "Showstoppers: Nine Reasons Why We Never Sent Our Special Operations Forces after Al Qaeda before 9/11," *The Weekly Standard* 9, no. 19 (2004): 25–33.
103. Available at: https://www.congress.gov/107/plaws/publ40/PLAW-107publ40.pdf.
104. Available at: https://www.nato.int/cps/en/natohq/topics_110496.htm.

105. Leigh Neville, *Special Forces in the War on Terror* (New York: Osprey Publishing, 2015), 22. See also Gary Schroen, *First In: An Insider's Account of How the CIA Spearheaded the War on Terror in Afghanistan* (San Francisco, CA: Presidio Press, 2005); and Gary Berntsen and Ralph Pezzullo, *Jawbreaker: The Attack on Bin Laden and Al-Qaeda* (New York: Three Rivers Press, 2005).

106. Kalev I. Sepp, Charles H. Briscoe, Richard L. Kiper, and James A. Schroder, *Weapon of Choice: ARSOF in Afghanistan* (Ft. Leavenworth, KS: Combat Studies Institute Press, 2003). See also Doug Stanton, *Horse Soldiers: The Extraordinary Story of a Band of U.S. Soldiers Who Rode to Victory in Afghanistan* (New York: Scribner, 2009).

107. Hy S. Rothstein, *Afghanistan and the Troubled Future of Unconventional Warfare* (Annapolis, MD: Naval Institute Press, 2006).

108. Details of this were told to the author by senior Army SOF leaders at Ft. Bragg in January 2011.

109. Available at: https://www.congress.gov/107/plaws/publ243/PLAW-107publ243.pdf.

110. Bob Woodward, *Plan of Attack* (New York: Simon & Schuster, 2004).

111. For an interesting discussion of how the AFO element arrived in Iraq by rental cars from Turkey, see Mark Giaconia, *Operation Viking Hammer: One Green Beret's Firsthand Account of Unconventional Warfare in Iraq, 2003* (Scotts Valley, CA: CreateSpace Independent Publishing Platform, 2018).

112. Woodward, *Plan of Attack*. See also Mike Tucker and Charles Faddis, *Operation Hotel California: The Clandestine War Inside Iraq* (Guilford, CT: The Lyons Press, 2008).

113. "In Brief—US Aid for Iraqi Opposition Groups." Available at: https://janes.ihs.com/Janes/Display/jdw03486-jdw-2003.

114. Available at: https://web.archive.org/web/20070208074448/http://www.cnn.com/2003/US/02/05/sprj.irq.powell.transcript/index.html.

115. "Iraq's Western Desert a 'Special Forces Playground.'" Available at: https://janes.ihs.com/Janes/Display/jdw04441-jdw-2003 (February 22, 2019).

116. Colonel Gregory Fontenot, USA (ret.), LTC E.J. Degen, USA, and LTC David Tohn, USA, *On Point: The United States Army in Operation IRAQI FREEDOM* (Ft. Leavenworth, KS: Combat Studies Institute Press, 2004), 402.

117. See Linda Robinson, *Masters of Chaos: The Secret History of the Special Forces* (New York: PublicAffairs, 2004); and Giaconia, *Operation Viking Hammer*.

118. Eliot A. Cohen, *Commandos and Politicians: Elite Military Units in Modern Democracies* (Cambridge, MA: Center for International Affairs, Harvard University, 1978).

Chapter 5

Irregular Soldiers and Rebellious States—Past, Present, Future

In this concluding chapter, the contemporary operating environment as well as potential future trends are discussed. Three key findings of the book are presented and six policy prescriptions are offered for consideration for improving the U.S. ability to execute effective indirect interventions. These capabilities are a useful and integral part of the American foreign and defense policy toolkit.

Both the Obama administration and the Trump administration entered office hoping to disengage the United States from the troubling COIN "forever wars" of Iraq and Afghanistan. President Obama, for his part, had hoped to pivot to Asia and he did pull out U.S. troops from Iraq in 2011. Nevertheless, then, the Islamic State appeared with its horrors in Syria and Iraq and we returned—albeit on a smaller scale. President Trump meanwhile talked about being tough on terrorism while also wanting to get out of Afghanistan and the Middle East. Pentagon planners talked about "great power competition" with China and Russia. The underlying assumptions of both administrations was that we needed to—although more likely we *wanted to*—return to a more traditional approach to militarily competing against other great power actors. But what if that is a chimera? The events following the Russian war with Georgia in 2008, and especially after the Crimea intervention in 2014, suggest that such GPC might not be about air wings, carrier battle groups, and armored divisions facing off against one another, but we may instead be heading back to the future, so to speak.[1] This is not to suggest that the United States should rest on its laurels about preparing for hi-tech war, but the tools addressed in this book might not be antiquated capabilities.

Sharing this view is the Princeton University Professor Michelle Lee. Writing in *Foreign Affairs*, she argued that conflict is now more likely to occur in the indirect domain through the proxy warfare method of "foreign subversion." In such warfare state-based actors use nonstate armed groups as their extensions. The proxies "inflict damage on target states with the aim of deconsolidating them and creating ungoverned space" in order to distract those states and deny them resources.[2] The logic of such subversion should be clear by now: why intervene directly when you can outsource the problem to others at a fraction of the human, economic, and political costs? Absent the imposition of heavy costs for behaving in such ways, it seems that such behavior may be more likely in the future.

This book has illustrated that the U.S. IC and military have had the history, tools, and individuals capable of both carrying out and defending against such subversion. However, the mainstream U.S. strategic and military cultures have been uncomfortable with such instruments. In Jan Gleiman's words, the mindset to be comfortable with such operating styles really is the "counter-culture" of the American way of war.[3] In this chapter we will discuss ways that the U.S. military can, or might be able to, cope with such competition if Lee and I, and others, are correct in our positions.

FID/UW AND THE AMERICAN APPROACH TO WARFARE

As the previous two chapters have shown, the U.S. IC and military can be effective at conducting both FID and UW campaigns. The cases studies presented here also show that such effectiveness is greatly increased when individuals are able to operate in depth in foreign cultures and are able to develop and build upon not only their knowledge of such regions but also their relationships. Like in many other areas of society, building one's network is important both professionally and personally. But the network is only truly optimized when one demonstrates their value substantively in terms of skills and abilities as well as socially. Few people will rate out as exceptional in both practical and social skills. For most it is about knowing one's shortcomings and trying to achieve some degree of balance there.

For members of the military this can be sometimes difficult because while some units might be more selective on certain attributes or skills, they are also created from, derived from, and their members come from initial training programs that inculcate the traditions and culture that the individual services have into individuals. They also demonstrate what their dominant professional ethos expects or demands from its members. Even for specialized units that select and assess based on aptitude in being able to learn languages and being mature enough to adapt to foreign cultures can run counter to their external image. SF, for instance, have recently been accused of being too focused on combat skills.[4] Some, such as Fletcher Schoen, have even argued that even SF's value of direct action skills and experience has diluted the language skills that might be needed for future UW, and by extension FID, missions.[5]

And a recent paper by SF colonel Edward C. Groot found that in a survey of over 1,200 Green Berets fully 26 percent of those soldiers believed that SF should focus on short-duration direct action missions with less focus on partnering and language skills.[6] Of course, one might counter with the argument that deployment operational tempos and individual personnel tempos for required skills and leadership training could factor into those issues. And such deployments likely may have improved particular language skills for some.

The SF have enacted two structural innovations that might favor the better employment of necessary special warfare skills for offensive and defensive indirect operations for both in the present and in the future. The first was the establishment of the 1st Special Forces Command (Airborne). This placed a major general in charge of the five active duty Special Forces Groups (1st, 3rd, 5th, 7th, and 10th), the two National Guard Special Forces Groups (19th and 20th), the two Psychological Operations Groups (4th and 8th), the 96th CA Brigade, and the 528th Sustainment Brigade. All of these units handle tasks and missions aligned with FID/UW, although they can also take part and assist with other missions such as direct action. It also created the 1st Special Operations Joint Task Force structure to provide a deployable, joint, scalable, and validated brigadier general-level command and control mechanism for special warfare missions. Such a dedicated special warfare command and control structure was needed in order to bridge the gap between U.S. unilateral CT efforts and the commitment of large-scale conventional forces. And these special warfare missions might include ultra-small conflicts or wars, low visibility or preparation of the environment missions, and the ability to enable regional or partner CT and security capabilities.[7]

The second innovation was the addition of the fourth battalion (sensitive activities) to the five active duty Special Forces Groups.[8] According to Army brigadier general Darsie Rogers, the purpose of these battalions are "to provide the Theater Special Operations Commanders with increased Special Warfare mission command capability, persistent preparation of the environment, and associated Special Warfare tasks through a purpose-built, regionally-expert force." These battalions include a Regional Support Element, Special Warfare Planning Detachment, and Jedburgh teams.[9] As each of the SF Groups is regionally aligned, such organizations should provide the regional COCOMs with more depth and expertise on special warfare topics. The Office of Special Warfare was also established at the U.S. Army Special Operations Command to support these efforts and to serve as a staff directorate, provide interagency linkages, and provide purposeful management, particularly for the conduct of UW.[10] As one senior Special Forces officer related to me, "The demand curve right now isn't for a Special Forces Sergeant fresh out of the [Special Forces Qualification Course]."[11]

But more work can be done. For instance, more emphasis should be placed in developing individual-level skills in cyber operations. As SF colonel Patrick Duggan has said, commanders tend to push such responsibilities onto signals or intelligence personnel when it should be an "everybody thing. . . . In the next 10 years, every single Green Beret, SEAL and Ranger must understand computers, cryptography and coding. It is essential to their survival. It is not a thing problem, it is a cultural one."[12] Such skills are also important in a day and age when adversaries use disinformation in strategic, operational, and tactical environments to sow division, angst, and distrust and one need to counter it. And, on the flipside, using such tools might even be able to alter the situations in a target country or region even without the introduction of U.S. personnel on the ground under certain circumstances.

THE CURRENT AND FUTURE OPERATING ENVIRONMENTS

While the future operating environment will surely contain differences from today due to societal, economic, military, and technological developments, it will probably also contain many similarities. The situations in Syria and Iraq over the past eight or nine years may be instructive. There the United States has been involved, on the one hand, in trying to assist Iraq defend its sovereignty while, on the other, conducting UW operations against the Islamic State across the border. The United States had faced considerable challenges in attempting the vetting, training, and equipping of "moderate" Free Syrian forces until Kurdish forces in the region became a more valuable proxy—if one that caused problems with our ostensible allies, the Turks.

While the United States has not been attempting regime change—at least not as a stated policy aim—against the Assad regime in Syria, it has been intervening within that state in order to combat and attrite the forces of the Islamic State whose numbers were swollen by an influx of transnational foreign fighters. In some ways this was similar to the situations in Nicaragua and El Salvador in the 1980s, while in other ways it was quite different—those countries did not share a land border, for instance. But the situation on the ground in Iraq has been quite difficult due to Iran's actions in supporting substate Shia militias in that country whose purposes are at odds with the U.S. objectives.

U.S. intelligence officers and military personnel will have to maintain maximum flexibility in being able to transition from one set of operating environment procedures to another. For the military in particular, they will

Table 5.1. U.S. Training of Syrian Opposition, 2011–2015[1]

Lead Agency	*CIA*	*Pentagon*
Dates	Vetting as of June 2012, training as of March 2013, funding cut in June 2015	Planned as of September 2014, Troop deployments and training February–July 2015, suspended October 2015
Training location	Jordan	Turkey, Arab partners
Number trained and equipped (goal)	10,000 (unknown)	180 (3,000 first year, 15,000 total)
Cost	Up to $1 billion/year, later reduced	Up to $500 million
Equipment	First nonlethal (communications equipment), then BGM-71 TOW missiles and small arms	Vehicles, electronics, and weapons

[1] The data in the table were compiled from newspaper sources Charles Edwards-Kuhn, April 16, 2016.

also have to be nimble enough to cope with the different sets of authorities that might be used in one area of operations, but not in the other. This will pose its own set of challenges as new operating TTP are used by adversaries.

And who will future adversaries be? The threat from Islamist extremists will still likely be present. Electoral politics will surely play a role in which threats are seen as most grave. But if American military history is any guide we probably will not be completely prescient about what threats might be seen as affecting vital or important national security interests in three, five, or ten years. Yes, clearly there are warnings and indicators about what might impact our interests, but how many predicted that we would intervene in Libya in 2011 even a year or two earlier?

It is likely that proxy actors will be utilized to counter U.S. interests. Such actors will take many forms, from hackers to criminal elements to terrorists (of many flavors and causes) to commercially motivated mercenaries. This last group is a big development over the past decade. While the United States contracted out to private security contractors for defensive, support, or logistical services in places like Afghanistan, Iraq, and Somalia, new groups like the Russian Wagner Group are of a somewhat unique flavor. Surely groups like the South African Executive Outcomes conducted offensive military operations for state or corporate clients, but Wagner has taken things to a whole new level by basically becoming a state-directed military force by cutout.[13] Such mercenary forces give governmental clients plausible deniability while allowing them to calibrate the use of force in order to achieve their policy ends. In addition, even if these ends are not achieved, the operational shortcomings of such hired units do not attach the same sorts of negative reputational or credibility effects that they would if carried out by the contracting nation's armed forces. And the lack of casualties to one's armed forces also can help to assuage domestic critics in a way that they would not possible when holding funerals with full military honors for one's own service members.[14]

As stated in the first chapter, Russia and China, and lesser powers like Iran and North Korea, will have their own interests that differ from ours. After nearly thirty years of witnessing the United States having difficulty in coping with threats below the level of major war, it seems safe to bet that those states and their proxies, pawns, and patsies will attempt to counter U.S. interests in the more messy political-military domains that lend themselves to indirect instruments of power. Surely a technological breakthrough in artificial intelligence or quantum computing might alter their calculus on such how to counter U.S. interests, but we are not there yet. (And yes, we must be innovative in these areas and should not let our guard down either.[15])

Largely falling in line with the above-mentioned items, the Army Special Operations Command in 2019 released a strategy document that argued that the trends in the global operating environment were pointing to global hyperconnectivity, environmental shifts, expanded competition space, and proliferation/access to WMD. Such hyperconnectivity will empower and favor those state and nonstate actors and individuals who have "*the agility and creativity to quickly exploit emerging capabilities and weaponize information.*" Environmental shifts, for their part, will affect social patterns, the climate, and economies and "*drive political instability and resource competition.*" Competition spaces will be expanded by rapid technological changes "*across the physical, virtual, and cognitive aspects of the environment.*" Last, the barriers to acquiring WMD technologies will continue to diminish.[16]

Factors such as pandemics (like the current COVID-19 outbreak) or climate change could have real geostrategic impacts by altering the productivity of economies and what governments choose to expend resources on. Information will likely continue to be weaponized. All of these factors will determine how the U.S. military and IC will be used. Declining defense and international affairs resources could cause U.S. foreign and defense policy shift. This could mean an increase in the use of small footprint indirect interventions or it could mean a reining in of the uses of military power.

KEY FINDINGS AND POLICY PRESCRIPTIONS

What should be the key takeaways from this book? I would suggest that there are three key findings from the analyses offered throughout.

Key Finding #1: Small Wars Are Not Going Away. As chapter 1 pointed out, the conduct of the so-called irregular operations has been the historical norm for the way the United States employs its military abroad. Across the 314 uses of the U.S. armed forces abroad from 1798 to 2018, fully 73.4 percent of these cases were for irregular operations. Even after the United States established itself as a superpower in 1941, the overwhelming way the armed forces were used was for such operations (64.8 percent of the cases). While it would be dangerous to make a straight-line projection that this will hold over time, it is safe to assume that this will continue to be the dominant mode of force employment for all of the reasons stated earlier. Both nation states and nonstate actors in the international system will likely try to avoid American conventional capabilities. The rise of "gray zone" threats from actors such as Russia will also mean that while we may try to reduce our involvement in such activities others may push for advantage in those areas.[17] Russia's Unit 29155, for instance, allegedly exists to conduct subversion, sabotage, and assassination activities across Europe in locations such as Moldova, Bulgaria, Montenegro, and the United Kingdom.[18]

Supporting allies, friends, neighbors, governments, or peoples in geopolitically key locations will likely mean that as much as we may try to avoid the messy ambiguity of such irregular operations, that will not always be possible. U.S. SOF have been working with friends and allies around the globe to counter GPC.[19] As Thomas Trask, Mark Clark, and Stuart Bradin have pointed out, "Both Russia and China have substantial holes in their territorial defense on the periphery, where SOF can present significant operational-level dilemmas that will detract from primary defenses."[20] And when such interventions are needed, public opinion considerations may mean that more indirect means of intervention may be more politically palatable. In addition, even in the application of large-scale "traditional" uses of force the indirect approaches argued here will not only be used but will be useful capabilities.

Key Finding #2: The Necessity of Building Expertise. The most successful examples of small-scale indirect interventions (El Salvador, OEF-P, Dhofar, and the assertive case in northern Iraq) showed that success was made possible due to personnel being present who not only were well trained for their tasks but had also prospered by having a depth of expertise on the languages and cultures of their operating areas and had developed the networks and relationships in order to maximize political-military effectiveness on the ground. Furthermore, some degree of repetition in deploying to areas, at least by some personnel, also seemed to make success more possible. One must understand not only the cultures of the people on the ground that one is assisting but also be able to understand and work with individuals in one's own other governmental agencies—for example, knowing how the embassy team operates and its key players and how to deal with competing agendas. Advisors that can operate in depth in their local environment, who know the cultural ins and outs and preferably the language, are more effective than those who are given limited training and preparation for such duties. They are also better able to know when they are trying to be used for purposes that might cause problems in the present or future.

While the dominant professional mindset of the military is that immersion in politics is a bad thing (and it certainly is domestically), overseas such understanding can be essential. No matter how compelling one's objectives might be, they will not be achievable if they will not work with the necessary individuals on the ground in their social and political system. One may have to accept suboptimal military achievements on the ground in the short term in order to ensure future operational success in one's area of operations. For this, military personnel on the ground need to have more of a fusionist perspective.[21]

Key Finding #3: Accept That Results Will Vary over Time. A key takeaway should be that while such indirect interventions may be able to achieve U.S. interests under certain circumstances, such success or failure is oftentimes of a perishable shelf-life. Success today can be failure tomorrow. And while hindsight may be 20/20, decision makers cannot always know what combination of variables will make today's success tomorrow's disaster. That is not to excuse strategic blunders or incompetence. Few today would have chosen to intervene in Iraq in 2003, given what we know now, for instance. I attempted to be as clearheaded as possible when assessing the nine historical examples presented here—warts and all. While we may disagree with the decision makers after the fact oftentimes, they have had access to information that they believe justifies—or justified—their choices. Discerning motives after the fact can be a fun exercise, but it can also cause a confirmation bias on the part of the second guesser due to information that decision makers did not possess at the time. Indirect interventions, be they offensive or defensive in nature, limited or assertive in scope, should be carefully considered. Circumstances on the ground are constantly shifting. There is no single template that will guarantee success; one must be vigilant of the circumstances on the ground and be wary of ideal types. But sometimes results are needed in a particular time and place. Uncertainty about future happenings cannot always place a brake on action.

What can be done to achieve objectives over time? I offer six prescriptions that may help to improve political-military effectiveness in the future.

Policy Prescription #1: Develop Considerations for What Mandates Indirect Interventions. Even for a superpower like the United States with a defense budget of $738 billion[22] and an international affairs budget of $56.6 billion,[23]

(4) Do the interests of the proxies align with U.S. interests and will we be able to sway operational and tactical decision-making of our partners on the ground?
(5) Will adverse actions by supported actors on the ground significantly increase negative perceptions of U.S. policy?
(6) Will an improper calibration of assistance cause more external intervention into the conflict? If so, what are the parameters for countering such external actors?

Answering these questions will not always be easy. Various domestic and foreign actors will have their own agendas. But at a minimum U.S. decision makers must answer these questions, and others, in order to make judicious decisions on the possibility of extending blood and treasure in such interventions.

Policy Prescription #2: Understand That the Concepts of War and Peace Are a False Dichotomy. No matter what direction American grand strategy goes in, U.S. policymakers must shunt aside the belief that conditions called "war" and "peace" exist in the twenty-first century.[36] America's adversaries in this disordered world certainly see no gain from thinking in terms of such binary choices. They likely do, however, get some comfort in knowing that many here see that distinction and operate with that mindset. Seeing things in such a binary fashion limits choices and reduces American flexibility in dealing with threats whether they are major or minor in scope. While oversight is very important in a democracy like the United States, their needs to be more allowance for how authorities are defined or else we risk creating seams in our capabilities that can be exploited by adversaries.

Policy Prescription #3: *Avoid Strategic Monism*. In his classic book *The Soldier and the State*, the late political scientist Samuel Huntington discussed the concepts of strategic monism and strategic pluralism. Strategic monists are those wedded to one unwavering strategic concept, whereas strategic pluralists are more heterodox in their views on the tools of strategy.[37] This book is not meant to offer indirect forms of intervention as the end-all-be-all concept that will achieve U.S. strategic objectives and offer a "cheap hawk" solution to strategic conundrums.[38] But on top of binning concepts like war and peace, leaders sometimes also bin concepts like favoring traditional military means over being able to operate across the domains of conflict. We need to be strategic pluralists in order to cope with the myriad threats of the twenty-first century.

Policy Prescription #4: Develop and Protect a Mindset for Irregular Operations. The fourth SOF truth is that "Competent Special Operations Forces cannot be created after emergencies occur."[39] This is also the case for leaders and individuals that are more comfortable operating in the murkier political-military milieu. Our experiences and results from both Iraq and Afghanistan show that political considerations are not something that can be thrown up on a PowerPoint slide but not be understood. All military leaders should be provided with some preparation in political-military affairs at the pre-accession level and such considerations should be addressed in formal professional military education at every level for company-grade, field-grade, and general (or flag-ranked) officers. Limited numbers of individuals with particular skills or promise in these areas should preferably be protected by promotion boards because the rigid hierarchies and career paths for military officers does not allow for lateral entry into the profession.[40] But the Army must also continue to refine the concept of special warfare.[41] Special warfare is a critical concept for the conduct of FID/UW but it is not enshrined in joint doctrine.[42]

For more specialized units used for indirect interventions, there should be units or individuals specifically assessed and selected based upon Anna Simons idea about the usefulness in possessing the ability for wile. For her such wile is the ability of individuals to: (1) identify the feature, or features in an adversary that can be used to permanently alter conditions; (2) know how to appeal to, outwit, or neutralize different audiences; (3) have an intuitive ability to test, tease, and probe so that you can make your own opportunities and are not operating on someone else's time line; and (4) have an appetite for turning the tables on adversaries in order to cause them to undo themselves.[43] This skillset is especially important in this day and age where adversaries prefer to operate through subversion. The Russians, Chinese, Iranians, and North Koreans, among others, are particularly skilled in political warfare. While they may be playing by a different set of rules, U.S. individuals with a keen understanding and ability for wile would be exceedingly helpful for operating in FID/UW environments.

Serious consideration should be given to establishing a special warfare-focused SMU devised of such individuals. This would be useful for developing a more robustly resourced unit with deeper expertise. Such expertise would likely channel back out to the non-SMU SF units just as the training and expertise of the 75th Ranger Regiment diffuses out to the U.S. Army and the Naval Special Warfare Development Group's expertise filters out to the other SEAL teams through personnel assignments. It would also perhaps bring a more balanced approach to JSOC. As Anna Simons has pointed out, "SOF's [direct action]-centric orientation has not resulted in the wholesale collapse of a single hostile nonstate entity. This (non-)result alone suggests the need to try something new."[44] Over time such a unit might also allow for a more indirect approach–focused individual to take command of SOCOM—as

JSOC is seen as a "kingmaker" (someday "queenmaker") assignment for selection as SOCOM commander.

Policy Prescription #5: Develop Coordination and Command and Control Mechanism for Indirect Interventions. Coordination and refined command and control elements will be essential for maximizing political-military effectiveness in indirect interventions. At the national level of coordination it is essential that the NSC should be used to harmonize policy across various government agencies and to referee squabbles.[45] This might be difficult at present due to the realities of the Trump administration's use of the NSC, but a positive fix would be to resurrect, under a new title, the Board for LIC there. Mandated by law under the Goldwater-Nichols Act and implemented by President Ronald Regan's NSDD-277 of June 15, 1987, this Board was instructed that it would address interagency responses to congressional queries on LIC and SOF and foreign assistance and intelligence requirements for LIC.[46] The Board was also to address how technology and competitive strategies would improve LIC strategies, how to improve communications and intelligence interoperability in LIC, how LIC structures would work with country and regional policy and coordination mechanisms, and whether adequate capabilities existed to execute the strategy elements of NSDD 277, and how to address shortfalls.[47] This would go a long way in ensuring oversight with Congress on issues such as the use of various authorities in the field.

Another possibility would be to re-establish the Eisenhower-era Planning Board in the NSC. The Planning Board, at least in theory,

> devises the policy options and longer-range plans presented to the NSC Principal's Committee and to the President, so that they could focus their attention on the big picture. Board members would invest enough time in interagency planning on behalf of the NSC to produce nuanced, far-sighted joint strategies, while also investing enough time in their home department to shape their Department's plans and execution.[48]

This might be an easier sell as it would not solely be focused upon the peculiarities of special warfare. And, of course, the particular personalities of the president and their National Security Advisor, and their type of leadership style, play crucial roles not only in decision-making but also in how the NSC staff is organized and used. Regardless, a strategic planning directorate on the NSC staff would help to coordinate interagency efforts and mitigate some problems.[49]

While the creation of the 1st Special Forces Command (Airborne) discussed earlier was a useful development, we should think bigger—even if a special warfare SMU is developed. A Joint Special Warfare Command commanded by an LTG or vice admiral that would be the equivalent of what the JSOC is for the surgical strike forces of the SMUs (i.e., Delta Force, Naval

Special Warfare Development Group, the Ranger Regiment, the 24th Special Tactics Squadron, and the Intelligence Support Activity) is long overdue. Sean Naylor has written about this idea.[50] And as the defense scholar Christopher Lamb has noted, "If SOF's indirect approach will be more important in the future, then USSOCOM needs to develop the same cross-organizational collaboration skills for the indirect SOF approach that it has used to such good effect in its direct approach to counterterrorism."[51] Such a command or subcommand thereof should also be the main effort and supported element in FID/UW environments.[52]

Military organizations, like most bureaucracies, are, as discussed earlier, very hidebound about judging themselves vis-à-vis other organizations about resources and rank structures. A lower rank structure and resourcing contributes to feelings that some organizations are better than others.[53] The fact that JSOC is composed of, as mentioned in the prologue, basically hyperconventional units that makes them more accepted by the conventional forces in the military and helps contribute to some looking down on the unique political-military challenges of the special warfare mission.

This is not to denigrate the capabilities of surgical strike–oriented SOF. They will remain key pieces for countering terrorism and conducting other critical assignments. But the recent Russian practices of using political warfare and UW in Ukraine, the situations in Syria and Iraq, and conflicts elsewhere show that special-warfare SOF supported and enabled by non-SOF military forces and other interagency partners will also be very busy for the foreseeable future.[54] Again, both are valuable instruments in the U.S. toolkit of capabilities but are best used when the circumstances dictate their use.

Policy Prescription #6: Develop and Expand Information and Cyber Capabilities for Indirect Interventions. Last, as was hinted at above, the United States needs to develop and expand its capabilities in developing and maintaining informational and cyber tools for indirect interventions. As one participant in the CNA conference stated:

> Today's high-stakes battles are being fought in the information environment—alternatively referred to as "the infospace" throughout the discussion. Social media campaigns (featuring widespread use of disinformation, "deep fake" images, false personas, and botnets), releases of hacked sensitive information, media censorship, and state-sponsored propaganda are replacing artillery, ships, and aircraft as the primary vectors of the daily competition for global influence. Information proliferates over a constantly expanding array of platforms and crosses the globe in a matter of seconds. Interjections from individuals (e.g., celebrities, influencers, thought leaders) and non-state actors compound the chaos of this crowded global conversation. The infospace is a virtually unregulated international meeting place where traditional linguistic and

> geospatial barriers deteriorate and the collective consciousness is vulnerable to influence and exploitation.[55]

In order to deal with this environment, I subscribe to that conference's findings that the United States must create an influence strategy for dealing with indirect interventions, invest in SOF's influence operations capabilities, embrace influence operations, and develop and maintain the requisite expertise.[56] In addition, a civilian counterpart organization, once a job of the U.S. Information Agency, should be considered to help align policy across the government. This might be doable through the State Department's Global Engagement Center,[57] but it might not be practical.[58]

Dealing with the informational and cyber domains is imperative because those domains will become more important over time. As one CNA conference participant noted, "Those of us who spent years studying Westphalia . . . [are] now peeling that back and realizing that a state is a collection of individuals with cellphones. That changes the concept of borders, satellites, constraints."[59] Working in the contemporary and future operating environments, and as seen in some of the cases in this book, will require the ability to influence and counter the influence of other actors.

CONCLUSION

This book has sought to offer a primer—or refresher—on how these forms of intervention work, discuss their key concepts, explain their promise and pitfalls, and provide useful examples of how defensive and offensive indirect interventions, be they limited or assertive, have been executed in across different operational eras and in different strategic contexts. It has also tried to explain them in terms of schools of military thought in the United States. One may agree or disagree with the findings and conclusions here. But even if one disagrees I hope that it will encourage some debate because whether one likes it or not, we will be grappling with many of the issues addressed here for the foreseeable future. In the end, the U.S. arsenal of "big stick" military tools must also be complemented through the maintenance of a smaller and more nimble stick that is both sharp and smart.

NOTES

1. One problem with assessing GPC is that the 2018 National Defense Strategy is a classified document with only eleven pages publicly released. Frank Hoffman, however, has noted that the 2018 NDS considers GPC threats across the spectrum of

conflict. See his "Sharpening Our Military Edge: The NDS and the Full Continuum of Conflict," *Small Wars Journal* (undated). Available at: https://smallwarsjournal.com/jrnl/art/sharpening-our-military-edge-nds-and-full-continuum-conflict.

2. Melissa M. Lee, "Subversive Statecraft: The Changing Face of Great-Power Conflict," *Foreign Affairs* (December 4, 2019). Available at: https://www.foreignaffairs.com/articles/2019-12-04/subversive-statecraft.

3. Jan K. Gleiman, "The American Counterculture of War: Supporting Foreign Insurgencies and the American Discourse of War," *Special Operations Journal* 1, no.1 (2015): 19–36.

4. *United States Special Operations Command Comprehensive Review* (Tampa, FL: USSOCOM, 23 January 2020). Available at: https://int.nyt.com/data/documenthelper/6736-special-operations-forces-review/c93cf96d67c341b2d2c3/optimized/full.pdf#page=1.

5. Fletcher Schoen, "Reorganization Is Imperative to Fixing Special Forces' Bent Unconventional Culture," *Small Wars Journal* (June 29, 2015). Available at: https://smallwarsjournal.com/jrnl/art/reorganization-is-imperative-to-fixing-special-forces%E2%80%99-bent-unconventional-culture.

6. Colonel Edward C. Croot, "There Is an Identity Crisis in Special Forces: Who Are the Green Berets Supposed to Be?" Army War College Fellows Strategy Research Project (2020), 34–40. Available at: https://warroom.armywarcollege.edu/wp-content/uploads/FSRP-AY20-USAWC-Fellow-COL-Ed-Croot-14-April-20.pdf.

7. BG Darsie Rogers, "1st Special Forces Command (Airborne) (Provisional)," briefing slides (undated). Available at: https://turcopolier.typepad.com/files/bgrogers.pdf. See also Joseph Trevithick, "The U.S. Army Has Quietly Created a New Commando Division," *Medium.com* (November 26, 2014). Available at: https://medium.com/war-is-boring/the-u-s-army-has-quietly-created-a-new-commando-division-2b90961b4821.

8. Eric P. Shwedo, "Expansibility and Army Special Operations Forces," *Parameters* (Winter 2017–18): 108–109.

9. Rogers, "1st Special Forces Command (Airborne) (Provisional)."

10. See *ibid.* and Douglas A. Livermore, "Passing the Paramilitary Torch from the CIA to Special Operations Command," *Military Times* (September 10, 2019). Available at: https://www.militarytimes.com/opinion/commentary/2019/09/10/passing-the-paramilitary-torch-from-the-cia-to-special-operations-command/.

11. Conversation with the author, September 9, 2017.

12. Duggan quoted in Todd South, "Cyber Skills Needed: Special Ops Leaders Seek Soldiers Who Can Fight the Enemy Up Close and Online," *Army Times* (October 3, 2017). Available at: https://www.armytimes.com/news/your-army/2017/10/03/hackers-wanted-special-ops-leaders-seek-soldiers-who-can-fight-the-enemy-up-close-and-online/.

13. For more on the role of mercenaries, see Sean McFate, *The New Rules of War: Victory in the Age of Durable Disorder* (New York: William Morrow, 2019), 115–142.

14. For similar arguments, see also Tor Bukkvoll and Åse G. Østensen, "The Emergence of Russian Private Military Companies: A New Tool of Clandestine Warfare," *Special Operations Journal* 6, no. 1 (2020): 1–17.

15. See, for example: Michael C. Horowitz, "Artificial Intelligence, International Competition, and the Balance of Power," *Texas National Security Review* 1, no.3 (May 2018): 37–57; and Michael Horowitz, Paul Scharre, Gregory C. Allen, Kara Frederick, Anthony Cho, and Edoardo Saravalle, "Artificial Intelligence and International Security," Washington, DC: Center for a New American Security (July 10, 2018). Available at: https://www.cnas.org/publications/reports/artificial-intelligence-and-international-security.

16. "Army Special Operations Forces Strategy" (Ft. Bragg, NC: U.S. Army Special Operations Command, 2019), 2. Available at: https://www.soc.mil/Assorted-Pages/ARSOF_Strategy.pdf.

17. "The purposeful application of multiple elements of power—information, economic, military, political—to achieve objectives in ways that exceed the threshold for normal competition yet fall below the level of major interstate war." National Intelligence Council, "Foreign Approaches to Gray Zone Conflicts" (March 2016). Available at: https://nsiteam.com/foreign-approaches-to-gray-zone-conflicts/. See also Hal Brands, "Paradoxes of the Gray Zone," *FPRI E-Note* (February 5, 2016). Available at: https://www.fpri.org/article/2016/02/paradoxes-gray-zone/; and Frank G. Hoffman, "Examining Complex Forms of Conflict: Gray Zone and Hybrid Challenges," *Prism* 7, no. 4 (2018): 30–47.

18. Michael Schwirtz, "Top Secret Russian Unit Seeks to Destabilize Europe, Security Officials Say," *The New York Times* (October 8, 2019).

19. See, for instance, Kyle Rempfer, "Looking for Friends: Green Berets Are Busy in China's Backyard," *Army Times* (July 12, 2019); and Eric Schmitt, "American Commandos Gear Up for New Shadow War with Russia," *The New York Times* (July 12, 2019).

20. See their "The Role of Special Operations Forces in a 'Great Power Conflict'," *MilitaryTimes.com* (May 4, 2020). Available at: https://www.militarytimes.com/opinion/commentary/2020/05/04/the-role-of-special-operations-forces-in-a-great-power-conflict/.

21. Sam C. Sarkesian, *Beyond the Battlefield: The New Military Professionalism* (New York: Pergamon Press, 1981), 172.

22. Joe Gould, "Pentagon Finally Gets Its 2020 Budget from Congress," *Army Times* (December 19, 2019). Available at: https://www.defensenews.com/congress/2019/12/19/pentagon-finally-gets-its-2020-budget-from-congress/.

23. Available at: https://www.usglc.org/the-budget/congress-finalizes-fy20-spending-international-affairs-budget-sees-small-increase-as-global-challenges-grow/.

24. William F. Weschler, "Counterterrorism by Proxy," *The National Interest* (March/April 2017): 26–27.

25. Claire Graja, rapporteur, *SOF and the Future of Global Competition* (Arlington, VA: CNA, May 2019), 12–13.

26. Tyrone L. Groh, *Proxy War: The Least Bad Option* (Palo Alto, CA: Stanford University Press, 2019), 203.

27. *Ibid.*, 202.

28. *Ibid.*, 210.

29. *Ibid.*, 202.

30. Mara E. Karlin, *Building Militaries in Fragile States: Challenges for the United States* (Philadelphia: University of Pennsylvania Press, 2018), 194.

31. *Ibid.*

32. *Ibid.*, 199–200.

33. *Ibid.*, 200.

34. *Counter-Unconventional Warfare White Paper* (Ft. Bragg, NC: U.S. Army Special Operations Command, 26 September 2014).

35. Stephen Biddle, Julia Macdonald, and Ryan Baker, "Small Footprint, Small Payoff: The Military Effectiveness of Security Force Assistance," *Journal of Strategic Studies* 41, no. 1–2 (2018): 132.

36. On this see, Adam Elkus and Michael P. Noonan, "Competitive Shaping in World Politics: A Bibliographic Essay and Course Outline," *The Philadelphia Papers* #18 (Philadelphia, PA: Foreign Policy Research Institute, May 2018).

37. Samuel P. Huntington, *The Soldier and the State: The Theory and Politics of Civil-Military Relations* (Cambridge, MA: Harvard University Press, 1957), 400 and 418–27.

38. Harvey Sicherman, "Cheap Hawks, Cheap Doves, and the Pursuit of Strategy," *Orbis* 49 (Autumn 2005): 613–629.

39. "Creation of competent, fully mission capable units takes time. Employment of fully capable special operations capability on short notice requires highly trained and constantly available SOF units in peacetime." Of course, I would eliminate the mention of peacetime. Available at: https://www.socom.mil/socpac/Pages/SOF-Truths.aspx.

40. The late Sam Sarkesian put it best when he argued that "the intellectual perspective to understand and appreciate the inter-penetration between military and civilian elements, and the political and social consequences of military intervention cannot be achieved by theoretical studies on a one-time basis. This can best be acquired by a grasp of the tools of political-social analysis. Thus, from the time of planning, through execution and aftermath, the military professional will find it extremely difficult to limit himself solely to the military element. He must be able to assess the political-military and social inter-relationships, and their consequences on his own military efficiency. Such knowledge is developed over a period of time and requires serious study with intellectual tools that are rarely part of traditional military education and training." Sarkesian, *Beyond the Battlefield*, 172.

41. "Special warfare is the execution of activities that involve a combination of lethal and nonlethal actions taken by a specially trained and educated force that has a deep understanding of cultures and foreign language, proficiency in small-unit tactics, and the ability to build and fight alongside indigenous combat formations in a permissive, uncertain, or hostile environment." *ADRP 3–05 Special Operations* (Washington, DC: Headquarters, Department of the Army, January 2018), 1–8.

42. Homer W. Harkins, "The Resurgence of Special Warfare," in *Special Operations: Out of the Shadows*, edited by Christopher Marsh, James D. Kiras, and Patricia J. Blockstone (Boulder, CO: Lynne Rienner, 2020), 59.

43. Anna Simons, *21st Century Cultures of War: Advantage Them* (Philadelphia Paper No. 1, Foreign Policy Research Institute, April 2013), 38.

44. Anna Simons, "Diversity and SOF: Boon or Bane?" *Special Operations Journal* 5, no. 1 (2019): 50.
45. John Gans, *White House Warriors: How the National Security Council Transformed the American Way of War* (New York: Liveright, 2019).
46. Available at: http://www.reagan.utexas.edu/archives/reference/Scanned%20 NSDDS/NSDD277.pdf.
47. Michael Noonan, "The Seductiveness of Special Ops?" *War on the Rocks* (March 3, 2015). Available at: https://warontherocks.com/2015/03/the-seductiveness-of-special-ops/.
48. F. G. Hoffman and Ryan Neuhard, "Avoiding Strategic Inertia: Enabling the National Security Council," *Orbis* 60, no. 2 (Spring 2016): 224.
49. Colin Dueck, "Strategic Planning for the Next President," *Orbis* 60, no. 3 (Summer 2016): 331–352.
50. Sean D. Naylor, "Support Grows for Standing Up an Unconventional Warfare Command," *Armed Forces Journal* (September 1, 2007). Available at: http://armed forcesjournal.com/support-grows-for-standing-up-an-unconventional-warfare-command/.
51. Christopher Lamb, "Global SOF and Interagency Collaboration," *Journal of Strategic Security* 7, no. 2 (Summer 2014): 8–20.
52. David J. O'Hearn, Damon S. Robins, and Aaron C. Sessoms, "Flattening the Learning Curve: SOF as the Supported Command in the Irregular Warfare Environment" (Master's Thesis, Naval Postgraduate School, June 2012).
53. Noonan, *The Seductiveness of Special Ops?*
54. Available at: https://www.socom.mil/socpac/Pages/SOF-Truths.aspx.
55. Graja, *SOF and the Future of Global Competition*, 3.
56. *Ibid.*, 5–6.
57. Available at: https://www.state.gov/bureaus-offices/under-secretary-for-public-diplomacy-and-public-affairs/global-engagement-center/.
58. Graja, *SOF and the Future of Global Competition*, 7.
59. *Ibid.*, 13.

Appendix A

U.S. "Irregular" and "Traditional" Operations Abroad, 1798–2018

Table A.1. U.S. "Irregular" Operations Abroad, 1798–2018

Small Scale (1–39,999 Troops)					*Large Scale (40,000+ Troops)*
Short Term (< 2 years)				*Long Term (> 2 years)*	*Duration*
Algiers, 1815	Panama (Colon),	China, 1934	Libya, 2012	Tripoli, 1801–1805	***Short Term***
Tripoli 1815	1885	Palestine, 1948	Southern Philippines,	Gulf of Mexico,	(< 2 years)
Spanish Florida—first	Panama, 1885	China, 1948–1949	2012	1806–1810	China, 1945
Seminole War,	Korea, 1888	China, 1954–1955	Niger, 2013	Caribbean, 1814–1825	Iraq (Provide
1816–1818	Samoa, 1888–1889	Egypt, 1956	Jordan, 2013	Africa, 1820–1823	Comfort), 1991
Amelia Island, 1817	Hawaiian Islands,	The Caribbean,	Philippines, 2013	Panama, 1903–1914	***Long Term***
Cuba, 1822	1889	1959–1960	Burundi, 2013	Cuba, 1906–1909	(> 2 years)
Cuba, 1823	Argentina, 1890	Congo, 1964	South Sudan,	Nicaragua, 1912–1925	Philippine Islands,
Cuba, 1824	Haiti, 1891	Dominican Republic,	2013–2014	China, 1912–1941	1899–1901
Puerto Rico (Spanish	Bering Strait, 1891	1965	Uganda, Democratic	Mexico, 1914–1917	Vietnam, 1964–1973
Territory), 1824	Chile, 1891	Congo, 1967	Republic of the	Haiti, 1915–1934	Bosnia, 1995–2008
Cuba, 1825	Hawaii, 1893	Cambodia, 1970	Congo, 2014	Dominican Republic,	Afghanistan,
Greece, 1827	Nicaragua, 1894	Evacuation from	Ukraine, 2014	1916–1922	2001–2018
Sumatra, 1832	China, 1894–1895	Cyprus, 1974	Liberia and Senegal,	Cuba, 1917–1922	Iraq, 2003–2011
Argentina, 1833	Korea. 1894–1896	Evacuation from	2014–2015	Nicaragua, 1926–1933	
Peru, 1835–1836	Columbia, 1895	Vietnam, 1975	Cameroon, 2015	Laos, 1962–1975	
Sumatra, 1838–1839	Nicaragua, 1896	Evacuation from	Yemen, 2016	El Salvador,	
Fiji, 1840	China, 1898–1899	Cambodia, 1975	Jordan, 2016	1981–1992	
Drummond Island,	Nicaragua, 1899	Lebanon, 1976	South Sudan, 2016	Sinai, 1982–2018	
Kingsmill Group, 1841	Samoa, 1899	Iran, 1980	Haiti, 2016–2017	Honduras, 1983–1989	
Samoa, 1841	China, 1900	Lebanon, 1982	Guatemala, 2017	Kosovo, 1999–2018	
China, 1843	Colombia (State of	Lebanon, 1982–1983	Honduras, 2017	East Timor, 1999–2002	
Africa, 1843	Panama), 1901	Italy, 1985	Peru, 2017	Philippines,	
Smyrna, 1849	Colombia, 1902	Bolivia, 1986	Uruguay, 2017	2002–2015	
Johanns Island (east of	Columbia (State of	Andean Initiative in	Sri Lanka, 2017	Pan Sahel, 2002–2018	
Africa), 1851	Panama), 1902	War on Drugs, 1989	Niger 2017	Yemen, 2002–2005	

Argentina, 1852–1853	Honduras, 1903	Liberia, 1990	Ecuador and Colombia, 2017	Djibouti, 2003–2018
Nicaragua, 1853	Dominican Republic, 1903	Zaire, 1991	Tanzania, 2018	Central Africa, 2011–2016
China, 1854	Syria, 1903	Sierra Leone, 1992	Panama, Colombia, Honduras, El Salvador, and Trinidad and Tobago, 2018	Iraq, 2014–2018
Nicaragua, 1854	Abyssinia, 1903–1904	Somalia, 1992–1993	Thailand, 2018	Syria, 2016–2018
China (Shanghai), 1854	Dominican Republic, 1904	Bosnia, 1993	Yemen, 2018	
China (Hong Kong), 1854	Panama, 1904	Somalia, 1993–1994		
Fiji Islands, 1855	Korea, 1904–1905	Rwanda, 1994		
Uruguay, 1855	Honduras, 1907	Haiti, 1994–1995		
Panama, 1856	Nicaragua, 1910	Somalia, 1995		
China, 1856	Honduras, 1911	Liberia, 1996		
Nicaragua. 1857	China 1911	Central African Republic, 1996		
Nicaragua, 1857	Honduras, 1912	Rwanda and Zaire, 1996		
Uruguay, 1858	Panama, 1912	Albania, 1997		
Fiji Islands, 1858	Cuba, 1912	Congo and Gabon, 1997		
Mexico, 1859	China, 1912	Sierra Leone, 1997		
China, 1859	Turkey, 1912	Cambodia, 1997		
Angola, Portuguese West	Mexico, 1913	Guinea-Bissau, 1998		
Africa, 1860	Haiti, 1914	Kenya and Tanzania, 1998		
Colombia (Bay of Panama), 1860	China, 1916	Albania, 1998		
Panama, 1865	China, 1917	Liberia, 1998		
China, 1866	Mexico, 1918–1919	Kenya, 1999		
Mexico, 1866	Panama, 1918–1920	Sierra Leone, 2000		
Nicaragua, 1867	Soviet Russia, 1918–1920	Yemen, 2000		
Formosa, 1867	Dalmatia, 1919	Georgia, 2002–2004		

(Continued)

Table A.1. (Continued)

Small Scale (1–39,999 Troops)				*Large Scale (40,000+ Troops)*
Short Term (< 2 years)			*Long Term (> 2 years)*	*Duration*
Japan (Osaka,	Turkey, 1919	Cote d'Ivoire, 2002		
Hiolo, Nagasaki,	Honduras, 1919	Liberia, 2003		
Yokohama, and	China, 1920	Mauritania, 2003		
Negata), 1868	Guatemala, 1920	Liberia, 2003		
Uruguay, 1868	Russia (Siberia),	Haiti, 2004		
Colombia, 1868	1920–1922	Haiti, 2004		
Mexico, 1870	Turkey, 1922	Kenya, 2004–2005		
Korea, 1871	China, 1922–1923	Ethiopia, 2004–2005		
Colombia (Bay of	Honduras, 1924	Eritrea, 2004–2005		
Panama), 1873	China, 1924	Lebanon, 2006		
Mexico, 1873–75	China 1925	Egypt, 2011		
Hawaiian Islands,	Honduras, 1925	Somalia, 2012		
1874	Panama, 1925	Yemen, 2012		
Mexico, 1876	China, 1926			
Egypt, 1882	China, 1927			
	China, 1932			

Table A.2. U.S. "Traditional" Operations Abroad, 1798–2018

Small Scale (1–39,999 Troops)				*Large Scale (40,000+ Troops)*
Short Term (< 2 years)			*Long Term (> 2 years)*	*Duration*
Mexico (Spain, 1806	Lebanon, 1958	Poland, 2016–2017	Undeclared Naval War with France, 1798–1801	***Short Term***
West Florida (Spain), 1810	Thailand, 1962	Korea, 2017	Formosa (Taiwan), 1950–1955	(< 2 years)
Amelia Island and other parts of eastern Florida (Spain), 1812	Cuba, 1962	Romania, 2017	Iraq, 1991–2003	Spanish-American War, 1898
West Florida (Spanish territory), 1813	Mayaguez incident, 1975	Greece, 2017		Desert Shield/Desert Storm, 1990–1991
Spanish Florida, 1816	South Korea, 1976	Kuwait, 2017		***Long Term***
Oregon, 1818	Zaire, 1978	Ukraine, 2018		(> 2 years)
Falkland Islands, 1831	Libya, 1981	Iceland, 2018		War of 1812, 1812–1815
Mexico, 1836	Egypt, 1983			Mexican War, 1846–1848
Mexico, 1842	Chad, 1983			Civil War, 1861–1865
Turkey, 1851	Grenada, 1983			World War I, 1917–1918
Japan, 1853–1854	Persian Gulf, 1984			World War II, 1941–1945
Ryukyu and Bonin Islands, 1853–1854	Libya, 1986			Korean War, 1950–1953
Turkey, 1858–1859	Libya, 1986			
Paraguay, 1859	Persian Gulf, 1987–1988			
Japan, 1863	Panama, 1988			
Japan, 1864	Libya, 1989			
Japan, 1864	Panama, 1989			
Haiti, 1888	Philippines, 1989			
Brazil, 1894	Panama, 1989–1990			
Tangier, Morocco, 1904	Kuwait, 1992			
Dominican Republic, 1914	Kuwait, 1993			
	Bosnia, 1993–1995			
	Iraq, 1993			
	Macedonia, 1993–1994			
	Haiti, 1993–1994			

(Continued)

Table A.2. (Continued)

Small Scale (1–39,999 Troops)			*Large Scale (40,000+ Troops)*
Short Term (< 2 years)		*Long Term (> 2 years)*	*Duration*
Panama-Costa Rica, 1921	Bosnia, 1994		
Cuba, 1932	Bosnia, 1995		
Newfoundland, Bermuda, St. Lucia, Bahamas, Jamaica, Antigua, Trinidad, and British Guiana, 1940	Afghanistan and Sudan, 1998		
	Iraq, 1998		
	Yugoslavia, 1999		
Greenland, 1941	Iraq, 1999–2000		
Dutch Guiana, 1941	Iraq, 2001		
Iceland, 1941	Libya, 2011		
Germany, 1941	Poland, 2014		
Trieste, 1946	Lithuania, 2014		
Berlin, 1948–1949	South Korea, 2014–2015		
	Kuwait, 2015		

Appendix B

Combatant Command Authorities and U.S. Special Operations Command as a Unified Combatant Command

"10 USC § 162 — Combatant commands: assigned forces; chain of command," accessible at: http://www.law.cornell.edu/uscode/text/10/162.

(a) Assignment of Forces.—

(1) As directed by the Secretary of Defense, the Secretaries of the military departments shall assign specified forces under their jurisdiction to unified and specified combatant commands or to the United States element of the North American Aerospace Defense Command to perform missions assigned to those commands. The Secretary of Defense shall ensure that such assignments are consistent with the force structure prescribed by the President for each combatant command.

(2) A force not assigned to a combatant command or to the United States element of the North American Aerospace Defense Command under paragraph (1) shall remain assigned to the military department concerned for carrying out the responsibilities of the Secretary of the military department concerned as specified in section 7013, 8013, or 9013 of this title, as applicable.

(3) A force assigned to a combatant command or to the United States element of the North American Aerospace Defense Command under this section may be transferred from the command to which it is assigned only—

(A) by authority of the Secretary of Defense; and

(B) under procedures prescribed by the Secretary and approved by the President.

(4) Except as otherwise directed by the Secretary of Defense, all forces assigned to a unified combatant command shall be under the command of the commander of that command. The preceding sentence applies to forces assigned to a specified combatant command only as prescribed by the Secretary of Defense.

(b) Chain of Command.—Unless otherwise directed by the President, the chain of command to a unified or specified combatant command runs—

(1) from the President to the Secretary of Defense; and
(2) from the Secretary of Defense to the commander of the combatant command.

"10 USC § 164 — Commanders of combatant commands: assignment; powers and duties," accessible at: http://www.law.cornell.edu/uscode/text/10/164.

(a) Assignment as Combatant Commander.—

(1) The President may assign an officer to serve as the commander of a unified or specified combatant command only if the officer—

(A) has the joint specialty under section 661 of this title; and
(B) has completed a full tour of duty in a joint duty assignment (as defined in section 664(d) of this title) as a general or flag officer.

(2) The President may waive paragraph (1) in the case of an officer if the President determines that such action is necessary in the national interest.

(b) Responsibilities of Combatant Commanders.—

(1) The commander of a combatant command is responsible to the President and to the Secretary of Defense for the performance of missions assigned to that command by the President or by the Secretary with the approval of the President.
(2) Subject to the direction of the President, the commander of a combatant command—

(A) performs his duties under the authority, direction, and control of the Secretary of Defense; and
(B) is directly responsible to the Secretary for the preparedness of the command to carry out missions assigned to the command.

(3) Among the full range of command responsibilities specified in subsection (c) and as provided for in section 161 of this title, the primary duties of the commander of a combatant command shall be as follows:

(A) To produce plans for the employment of the armed forces to execute national defense strategies and respond to significant military contingencies.
(B) To take actions, as necessary, to deter conflict.
(C) To command United States armed forces as directed by the Secretary and approved by the President.

(c) Command Authority of Combatant Commanders.—

(1) Unless otherwise directed by the President or the Secretary of Defense, the authority, direction, and control of the commander of a combatant command with respect to the commands and forces assigned to that command include the command functions of—

(A) giving authoritative direction to subordinate commands and forces necessary to carry out missions assigned to the command, including authoritative direction over all aspects of military operations, joint training, and logistics;
(B) prescribing the chain of command to the commands and forces within the command;
(C) organizing commands and forces within that command as he considers necessary to carry out missions assigned to the command;
(D) employing forces within that command as he considers necessary to carry out missions assigned to the command;
(E) assigning command functions to subordinate commanders;
(F) coordinating and approving those aspects of administration and support (including control of resources and equipment, internal organization, and training) and discipline necessary to carry out missions assigned to the command; and
(G) exercising the authority with respect to selecting subordinate commanders, selecting combatant command staff, suspending subordinates, and convening courts-martial, as provided in subsections (e), (f), and (g) of this section and section 822(a) of this title, respectively.

(2)

(A) The Secretary of Defense shall ensure that a commander of a combatant command has sufficient authority, direction, and control over the commands and forces assigned to the command to exercise effective command over those commands and forces. In carrying out this subparagraph, the Secretary shall consult with the Chairman of the Joint Chiefs of Staff.

(B) The Secretary shall periodically review and, after consultation with the Secretaries of the military departments, the Chairman of the Joint Chiefs of Staff, and the commander of the combatant command, assign authority to the commander of the combatant command for those aspects of administration and support that the Secretary considers necessary to carry out missions assigned to the command.

(3) If a commander of a combatant command at any time considers his authority, direction, or control with respect to any of the commands or forces assigned to the command to be insufficient to command effectively, the commander shall promptly inform the Secretary of Defense.

(d) Authority Over Subordinate Commanders.—Unless otherwise directed by the President or the Secretary of Defense—

(1) commanders of commands and forces assigned to a combatant command are under the authority, direction, and control of, and are responsible to, the commander of the combatant command on all matters for which the commander of the combatant command has been assigned authority under subsection (c);
(2) the commander of a command or force referred to in clause (1) shall communicate with other elements of the Department of Defense on any matter for which the commander of the combatant command has been assigned authority under subsection (c) in accordance with procedures, if any, established by the commander of the combatant command;
(3) other elements of the Department of Defense shall communicate with the commander of a command or force referred to in clause (1) on any matter for which the commander of the combatant command has been assigned authority under subsection (c) in accordance with procedures, if any, established by the commander of the combatant command; and
(4) if directed by the commander of the combatant command, the commander of a command or force referred to in clause (1) shall advise the commander of the combatant command of all communications to and from other elements of the Department of Defense on any matter for which the commander of the combatant command has not been assigned authority under subsection (c).

(e) Selection of Subordinate Commanders.—

(1) An officer may be assigned to a position as the commander of a command directly subordinate to the commander of a combatant command or, in the case of such a position that is designated under section 601 of this title as a position of importance and responsibility, may be recommended to the President for assignment to that position, only—

(A) with the concurrence of the commander of the combatant command; and
(B) in accordance with procedures established by the Secretary of Defense.

(2) The Secretary of Defense may waive the requirement under paragraph (1) for the concurrence of the commander of a combatant command with regard to the assignment (or recommendation for assignment) of a particular officer if the Secretary of Defense determines that such action is in the national interest.

(3) The commander of a combatant command shall—

(A) evaluate the duty performance of each commander of a command directly subordinate to the commander of such combatant command; and
(B) submit the evaluation to the Secretary of the military department concerned and the Chairman of the Joint Chiefs of Staff.

(4) At least one deputy commander of the combatant command the geographic area of responsibility of which includes the United States shall be a qualified officer of a reserve component of the armed forces who is eligible for promotion to the grade of O–9, unless a reserve component officer is serving as commander of that combatant command.

(f) Combatant Command Staff.—

(1) Each unified and specified combatant command shall have a staff to assist the commander of the command in carrying out his responsibilities. Positions of responsibility on the combatant command staff shall be filled by officers from each of the armed forces having significant forces assigned to the command.
(2) An officer may be assigned to a position on the staff of a combatant command or, in the case of such a position that is designated under section 601 of this title as a position of importance and responsibility, may be recommended to the President for assignment to that position, only—

(A) with the concurrence of the commander of such command; and
(B) in accordance with procedures established by the Secretary of Defense.

(3) The Secretary of Defense may waive the requirement under paragraph (2) for the concurrence of the commander of a combatant command with regard to the assignment (or recommendation for assignment) of a particular officer to serve on the staff of the combatant command if the Secretary of Defense determines that such action is in the national interest.

(g) Authority to Suspend Subordinates.—

In accordance with procedures established by the Secretary of Defense, the commander of a combatant command may suspend from duty and recommend the reassignment of any officer assigned to such combatant command.

(h) Support to Chairman of the Joint Chiefs of Staff.—

The commander of a combatant command shall provide such information to the Chairman of the Joint Chiefs of Staff as may be necessary for the Chairman to perform the duties of the Chairman under section 153 of this title.

"10 U.S. Code § 167. Unified combatant command for special operations forces," accessible at: https://www.law.cornell.edu/uscode/text/10/167.

(a) Establishment.—

With the advice and assistance of the Chairman of the Joint Chiefs of Staff, the President, through the Secretary of Defense, shall establish under section 161 of this title a unified combatant command for special operations forces (hereinafter in this section referred to as the "special operations command"). The principal function of the command is to prepare special operations forces to carry out assigned missions.

(b) Assignment of Forces.—

Unless otherwise directed by the Secretary of Defense, all active and reserve special operations forces of the armed forces stationed in the United States shall be assigned to the special operations command.

(c) Grade of Commander.—

The commander of the special operations command shall hold the grade of general or, in the case of an officer of the Navy, admiral while serving in that position, without vacating his permanent grade. The commander of such command shall be appointed to that grade by the President, by and with the advice and consent of the Senate, for service in that position.

(d) Command of Activity or Mission.—

(1) Unless otherwise directed by the President or the Secretary of Defense, a special operations activity or mission shall be conducted under the command of the commander of the unified combatant command in whose geographic area the activity or mission is to be conducted.

(2) The commander of the special operations command shall exercise command of a selected special operations mission if directed to do so by the President or the Secretary of Defense.

(e) Authority of Combatant Commander.—

(1) In addition to the authority prescribed in section 164(c) of this title, the commander of the special operations command shall be responsible for, and shall have the authority to conduct, all affairs of such command relating to special operations activities.

(2) Subject to the authority, direction, and control of the Assistant Secretary of Defense for Special Operations and Low Intensity Conflict, the commander of such command shall be responsible for, and shall have the authority to conduct, the following functions relating to special operations activities (whether or not relating to the special operations command):

(A) Developing strategy, doctrine, and tactics.

(B) Preparing and submitting to the Secretary of Defense program recommendations and budget proposals for special operations forces and for other forces assigned to the special operations command.

(C) Exercising authority, direction, and control over the expenditure of funds—

(i) for forces assigned to the special operations command; and

(ii) for special operations forces assigned to unified combatant commands other than the special operations command, with respect to all matters covered by paragraph (4) and, with respect to a matter not covered by paragraph (4), to the extent directed by the Secretary of Defense.

(D) Training assigned forces.

(E) Conducting specialized courses of instruction for commissioned and noncommissioned officers.

(F) Validating requirements.

(G) Establishing priorities for requirements.

(H) Ensuring the interoperability of equipment and forces.

(I) Formulating and submitting requirements for intelligence support.
(J) Monitoring the promotions of special operations forces and coordinating with the military departments regarding the assignment, retention, training, professional military education, and special and incentive pays of special operations forces.

(3) The commander of the special operations command shall be responsible for—

(A) ensuring the combat readiness of forces assigned to the special operations command; and
(B) monitoring the preparedness to carry out assigned missions of special operations forces assigned to unified combatant commands other than the special operations command.

(4)

(A) The commander of the special operations command shall be responsible for, and shall have the authority to conduct, the following:

(i) Development and acquisition of special operations-peculiar equipment.
(ii) Acquisition of special operations-peculiar material, supplies, and services.

(B) Subject to the authority, direction, and control of the Secretary of Defense, the commander of the command, in carrying out his functions under subparagraph (A), shall have authority to exercise the functions of the head of an agency under chapter 137 of this title.

(C)

(i) The staff of the commander shall include a command acquisition executive, who shall be responsible for the overall supervision of acquisition matters for the special operations command. The command acquisition executive shall have the authority to—

(I) negotiate memoranda of agreement with the military departments to carry out the acquisition of equipment, material, supplies, and services described in subparagraph (A) on behalf of the command;
(II) supervise the acquisition of equipment, material, supplies, and services described in subparagraph (A),

regardless of whether such acquisition is carried out by the command, or by a military department pursuant to a delegation of authority by the command;

(III) represent the command in discussions with the military departments regarding acquisition programs for which the command is a customer; and

(IV) work with the military departments to ensure that the command is appropriately represented in any joint working group or integrated product team regarding acquisition programs for which the command is a customer.

(ii) The command acquisition executive of the special operations command shall be responsible to the commander for rapidly delivering acquisition solutions to meet validated special operations-peculiar requirements, subordinate to the Defense Acquisition Executive in matters of acquisition, subject to the same oversight as the service acquisition executives, and included on the distribution list for acquisition directives and instructions of the Department of Defense.

(D) The staff of the commander shall include an inspector general who shall conduct internal audits and inspections of purchasing and contracting actions through the special operations command and such other inspector general functions as may be assigned.

(f) Administrative Chain of Command.—

(1) Unless otherwise directed by the President, the administrative chain of command to the special operations command runs—

(A) from the President to the Secretary of Defense;

(B) from the Secretary of Defense to the Assistant Secretary of Defense for Special Operations and Low Intensity Conflict; and

(C) from the Assistant Secretary of Defense for Special Operations and Low Intensity Conflict to the commander of the special operations command.

(2) For purposes of this subsection, administrative chain of command refers to the exercise of authority, direction and control with respect to the special operations-peculiar administration and support of the special operations command, including the readiness and organization of special operations forces, resources and equipment, and civilian personnel. It does not refer to the exercise of authority, direction, and control of operational matters that are subject to the operational chain of command of the commanders of combatant commands or the exercise of authority, direction, and control of personnel, resources, equipment, and other matters that are not special operations-peculiar that are the purview of the armed forces.

(g) Budget.—In addition to the activities of a combatant command for which funding may be requested under section 166(b) of this title, the budget proposal of the special operations command shall include requests for funding for—

(1) development and acquisition of special operations-peculiar equipment; and
(2) acquisition of other material, supplies, or services that are peculiar to special operations activities.

(h) Intelligence and Special Activities.—

This section does not constitute authority to conduct any activity which, if carried out as an intelligence activity by the Department of Defense, would require a notice to the Select Committee on Intelligence of the Senate and the Permanent Select Committee on Intelligence of the House of Representatives under title V of the National Security Act of 1947 (50 U.S.C. 3091 et seq.).

(i) Regulations.—

The Secretary of Defense shall prescribe regulations for the activities of the special operations command. Such regulations shall include authorization for the commander of such command to provide for operational security of special operations forces and activities.

(j) Identification of Special Operations Forces.—

(1) Subject to paragraph (2), for the purposes of this section special operations forces are those forces of the armed forces that—

(A) are identified as core forces or as augmenting forces in the Joint Chiefs of Staff Joint Strategic Capabilities Plan, Annex E, dated December 17, 1985;
(B) are described in the Terms of Reference and Conceptual Operations Plan for the Joint Special Operations Command, as in effect on April 1, 1986; or
(C) are designated as special operations forces by the Secretary of Defense.

(2) The Secretary of Defense, after consulting with the Chairman of the Joint Chiefs of Staff and the commander of the special operations command, may direct that any force included within the description in paragraph (1)(A) or (1)(B) shall not be considered as a special operations force for the purposes of this section.

(k) Special Operations Activities.—For purposes of this section, special operations activities include each of the following insofar as it relates to special operations:

(1) Direct action.
(2) Strategic reconnaissance.
(3) Unconventional warfare.
(4) Foreign internal defense.
(5) Civil affairs.
(6) Military information support operations.
(7) Counterterrorism.
(8) Humanitarian assistance.
(9) Theater search and rescue.
(10) Such other activities as may be specified by the President or the Secretary of Defense.

(l) Budget Support for Reserve Elements.—

(1) Before the budget proposal for the special operations command for any fiscal year is submitted to the Secretary of Defense, the commander of the command shall consult with the Secretaries of the military departments concerning funding for reserve component special operations units. If the Secretary of a military department does not concur in the recommended level of funding with respect to any such unit that is under the jurisdiction of the Secretary, the commander shall include with the budget proposal submitted to the Secretary of Defense the views of the Secretary of the military department concerning such funding.

(2) Before the budget proposal for a military department for any fiscal year is submitted to the Secretary of Defense, the Secretary of that military department shall consult with the commander of the special operations command concerning funding for special operations forces in the military personnel budget for a reserve component in that military department. If the commander of that command does not concur in the recommended level of funding with respect to reserve component special operations units, the Secretary shall include with the budget proposal submitted to the Secretary of Defense the views of the commander of that command.

Index

Note: Page references for figures are italicized.

About the Author

Michael P. Noonan, PhD, is a senior fellow in the national security program at the Foreign Policy Research Institute. A veteran of Operation Iraqi Freedom, where he served as an advisor with an Iraqi infantry battalion, he has taught at Haverford College, Loyola University Chicago, and Swarthmore College. His writings have appeared in *The American Interest*, *Orbis*, *Parameters*, *National Security Studies Quarterly*, *War on the Rocks*, and U.S. News and World Report's *World Report*.

www.ingramcontent.com/pod-product-compliance
Ingram Content Group UK Ltd.
Pitfield, Milton Keynes, MK11 3LW, UK
UKHW012119150926
475560UK00002B/180

9 781442 271296